CASES IN CLIMATE CHANGE POLICY: POLITICAL REALITY IN THE EUROPEAN UNION

To Anneli and the lemurs of Madagascar
May their future be climate-change free

CASES IN CLIMATE CHANGE POLICY: POLITICAL REALITY IN THE EUROPEAN UNION

Edited by

Ute Collier and Ragnar E Löfstedt

Earthscan Publications Ltd, London

First published in the UK in 1997 by
Earthscan Publications Limited

Copyright © Ute Collier and Ragnar E Löfstedt, 1997

A catalogue record for this book is available from the British Library

ISBN: 1 85383 414 9

Typesetting and page design by PCS Mapping & DTP, Newcastle upon Tyne

Printed and bound in Great Britain by Clays Ltd, St Ives plc

Cover design by Yvonne Booth

For a full list of publications please contact

Earthscan Publications Limited
120 Pentonville Road
London N1 9JN
Tel: (0171) 278 0433
Fax: (0171) 278 1142
Email: earthinfo@earthscan.co.uk
WWW: http://www.earthscan.co.uk

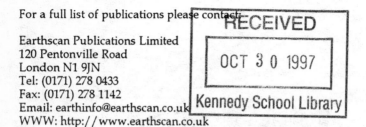

Earthscan is an editorially independent subsidiary of Kogan Page Limited and
publishes in association with WWF-UK and the International Institute for
Environment and Development.

CONTENTS

FOREWORD

Climate change represents one of the most difficult environmental problems with which policy makers have had to grapple. The production of greenhouse gases is deeply embedded in the way in which a modern society operates: in the way we heat our homes, use electric appliances, commute to work and travel during our leisure time. Greenhouse gas emissions also arise from the manufacture of the goods we consume. Climate change is quite unlike other environmental problems which could be addressed by targeting measures on a few large companies making up a sector such as chemicals or electricity supply. Addressing the problem of climate change requires coordinated action by consumers, business, governments and local authorities. Orchestrating this activity within and across nations represents an unprecedented challenge.

The publication of this book is timely. In mid–1996, the Intergovernmental Panel on Climate Change (IPCC) produced its Second Assessment Report, projecting that climate change would take place throughout the 21st century, and beginning to assess what the impacts of these changes would be across the globe. The second IPCC report confirmed what many suspected – that the recurrence of warmer years during the 1980s and 1990s was not simply a statistical aberration but that 'the balance of evidence suggests a discernible human influence on the global climate'.

At the same time, the developed nations of the world, among which the 15 European Union Member States figure prominently, are preparing to make commitments to constrain emissions of carbon dioxide (CO_2), the main greenhouse gas. The so-called 'Berlin Mandate', put in place at the first meeting of the Conference of the Parties to the Framework Convention on Climate Change (FCCC) in April 1995, requires developed countries to begin new talks on a new protocol imposing binding obligations to reduce CO_2 emissions beyond the year 2000.

Getting an agreement under the Berlin Mandate will solve only part of the problem – perhaps the easier part. Implementing policies to reduce greenhouse gas emissions is the major challenge. A number of developed countries, including some which project a 'green' image on the international stage, will fail to stabilise CO_2 emissions at their 1990 levels by the year 2000. This 'soft' commitment was built into the FCCC itself. The Netherlands, for example, developed a coherent and transparent climate programme which was widely studied by policy makers in other countries.

In 1996 the Netherlands was forced to introduce a short term 'crash' programme for greenhouse gas emissions reductions in order to prevent a planned emissions reduction by the year 2000 turning into an emissions increase.

Climate policy in the European Union and its Member States presents unique difficulties. In one sense, the EU is a microcosm of the global community of nations. The EU is attempting to coordinate the activities of a diverse set of nations in support of a common position. The EU Member States vary, not only in their level of economic development, but also in terms of culture, the priority attached to environmental protection, and the way that natural resources are marshalled to meet energy and transport needs.

Climate policy is being made against the background of other major developments within the EU. For example, the energy sector, a major source of greenhouse gases, is being challenged by the goal of market liberalisation. Electricity and gas utilities, which have until now enjoyed a secure monopoly position, will have to learn how to live in a world where their customers have the freedom to choose alternative suppliers. This has profound implications for the way in which markets, as well as policy makers, will determine patterns of activity, technology choice and hence environmental impacts. The availability and choice of policy levers which can be used to influence greenhouse gas emissions are significantly altered by energy market liberalisation. This difficult issue is tackled squarely in the pages of this book.

EU climate policy during the 1990s has developed in the context of a major political debate about the appropriate relationship between the Union and its members. The principle of subsidiarity – taking decisions as close to the public as possible – was introduced in the Maastricht Treaty. This has reinforced the role of Member States and weakened the capacity of the EU to make policy. The EU has been particularly unsuccessful in forging a common climate policy – partly because the Commission has relied heavily on fiscal measures, notably the infamous carbon/energy tax, concerning which some Member States have jealously guarded their sovereignty.

The difficulties of making climate policy at the EU level are partly those of environmental policy in general. Agreement on any particular measure is determined by a complex interplay between the Commission, the Council of Ministers and the European Parliament. Alongside the formal procedures, a flourishing sector of trade associations, company representatives and lobby groups is locating itself in Brussels in order to monitor – and influence – the development of EU measures. In practice, much of this influence is exercised even before proposals for new measures leave the Commission to be considered by the Council and Parliament. Like Washington DC, Brussels is developing its own policy-making dynamic – but with a distinctly European flavour. Like Washington DC, it is proving a fertile ground for lobbyists and academic observers alike.

Within this chaotic policy-making milieu, forging a European climate policy has proved particularly difficult. Activities which lead to greenhouse gas emissions pervade all aspects of economic and social life. The climate change policy agenda has necessarily attracted the attention of a wide range of interest groups, many of which are threatened by climate policies. The consequent gridlock has left EU climate policy very much in the hands of the individual Member States – each of which has its own distinct approach and each of which faces its own complex set of challenges.

This is where the value of this book lies. The editors bring their own rich experience to this study of national climate policies. Both are familiar with the challenges of comparing and contrasting national approaches – and have a truly European perspective. The contributions to the book rest on a solid body of empirical research which provides an authoritative account of: what countries have been doing about greenhouse gas emissions; the struggles which took place to get these policies enacted; how climate policy has interacted with other developments; and the challenges and opportunities facing countries as they approach real climate policy commitments under the Berlin Mandate process. It also usefully explores some generic issues in climate policy – such as the link to market liberalisation – which have applicability across national boundaries. If there is to be a truly European dimension to climate policy, it will almost certainly have to take account of these more generic factors.

The 1992 Framework Convention on Climate Change set in train a complex, multi-layered process of policy development. Taking the model of previous international agreements, policy development is likely to stretch out over years, perhaps decades. Progress will almost certainly be much slower than those with a profound concern over climate change would like. This will certainly not be the last book written on the subject of climate policy. But it comes at a pivotal point, as evidence about the reality of climate change firms up and as developed countries begin to measure up to the challenge of implementing serious greenhouse gas reduction measures. This book provides a welcome review of the state-of-the-art at this time.

Jim Skea
Director
Global Environmental Change Programme
UK Economic and Social Research Council

PREFACE AND ACKNOWLEDGEMENTS

This book has its roots in a project entitled 'Climate Change Polices in the European Union', based at the Robert Schuman Centre of the European University Institute in Florence, Italy, for which generous cofunding was provided by the Directorate General for the Environment (DG XI) of the European Commission. The 18-month project under the direction of Professor Yves Mény was coordinated by Ute Collier, and involved an international set of collaborators. The main focus of the project was on the political feasibility of policy action related to climate change. The principle aim was to identify in a systematic fashion what can be broadly defined as political constraints which hinder the definition and the implementation of optimal climate change policies in the European Union (EU).

Work on the project took place between October 1994 and March 1996. It was decided to focus on the seven Member States (of which six are discussed in this book) considered to provide a representative picture of the diverse situations within the 15 nations. They were also to include the EU's largest emitters of CO_2, as well as a look at 'old' and 'new' Member States. Hence, Denmark (which is omitted from the book for space reasons), Germany, France, Italy, Spain, Sweden and the UK were chosen, whose combined CO_2 emissions account for nearly 85 per cent of EU's total. These countries have very different characteristics with respect to fuel use, make up of the electricity sector, energy intensity of the economy and sensibility to environmental problems, which are all factors that can be expected to shape climate change policies.

Some very detailed country case studies resulted from the project work, which hopefully will assist the Comission in its assessment of Member States' climate change policies. Considering the strong interest in the public domain and in academic circles, both in the climate change issues and in EU environmental policy, the project members decided to pursue the publication of an edited volume based on material from the project, as well as some additional background material. Earthscan was a first preference as publisher considering its excellent reputation in the environmental field and its support for this project is greatly appreciated. Ute Collier and Ragnar Löfstedt (one of the project collaborators) took in hand the editorial responsibilities. Publication of this volume was agreed to by the Commission but it must be stressed that the opinions expressed herein are the sole responsibility of the authors and do not in any way reflect the views of the Commission or Member State governments.

The book would not have been possible without the many people who provided us with references, commented on innumerable drafts and encouraged us to complete it. In particular, we are indebted to Tom Downing of the Environmental Change Unit at Oxford University, Michael Grubb from the Energy and Environment Programme at the Royal Institute of International Affairs, Laura Kelly at ActionAid, Gerald Leach at Stockholm Environment Institute and Evert Vedung at the Department of Government, University of Uppsala, for commenting on earlier versions of several chapters that appear in this book. A large thank you goes to Dean Anderson at the Energy and Environment Programme, Royal Institute of International Affairs, who read the draft in its entirety and provided us with many useful comments and suggestions. We are also grateful to Jim Skea from the Science and Policy Research Unit at the University of Sussex for writing an excellent Foreword. Furthermore, we would like to thank all the interviewees who willingly participated in this study, as well as the authors of the individual case studies who produced brilliant chapters, and yet met the tight deadlines we imposed on them and put up with our criticisms.

A particularly big thank you must go to Eric Palmer for spending two hot July weeks in Florence assisting with the final editing, spotting numerous inconsistencies and making language corrections, as well as adding conciseness to the conclusions. We are also grateful to our editor Jo O'Driscoll at Earthscan not only for her support and encouragement, but also for providing us with a much needed advance to help cover some of the coordination costs of this book. Furthermore, we are indebted to our respective 'bosses', Professors Yves Mény at the European University Institute and Roland Clift at the Centre for Environmental Strategy, for their support in making this book become a reality. Finally, Ute Collier would like to express her appreciation to Eric Palmer for his patience and putting up with the many moans about the book, while Ragnar Löfstedt would like to thank Laura Kelly for her continued love and support and to apologize to Anneli Löfstedt for neglecting his fatherly duties as the book neared completion.

Ute Collier and Ragnar E Löfstedt,
Florence and Guildford,
August 1996

CONTRIBUTORS

Ute Collier: Robert Schuman Centre, European University Institute, Via dei Roccentini 9, 50016 San Domenico di Fiesole (Fi), Italy

Pierre-Noël Giraud: CERNA – ENSMP, Ecole des Mines, 60, blvd St. Michel, 75272 Paris Cedex 06, France

Michael Huber: Institut für Soziologie, Universität Hamburg, Allende-Platz 1, 20146 Hamburg, Germany

Xavier Labandeira Villot: Departamento de Economía Aplicada, Universidade de Vigo, Apdo. 874, 36200 Vigo, Spain

Ragnar E Löfstedt: Centre for Environmental Strategy, University of Surrey, Guildford, Surrey, GU2 5XH, UK

Susan Owens: Department of Geography, University of Cambridge, Downing Place, Cambridge, CB2 3EN, UK

Gianni Silvestrini: Consiglio Nazionale delle Ricerche, Istituto per l'edilizia ed il risparmio energetico (CNR-IEREN), Via Cardinale M. Rampolla 8/D, 90142 Palermo, Italy

Jo Smith: Department of Geography, University of Cambridge, Downing Place, Cambridge, CB2 3EN, UK

LIST OF FIGURES

LIST OF TABLES

ABBREVIATIONS AND ACRONYMS

ABB	Asea Brown Boveri
ADEME	Agence de l'Environment et de la Maitrise de L'Energie (France)
AGR	Advanced Gas Cooled Reactor
ALTENER	An EU decision concerning renewable energy sources
AOSIS	Alliance of Small Island States
BDI	Bund Deutscher Industrie (association of German industry)
BMU	Bundesministerium für Umwelt (Germany)
CCGTs	Combined Cycle Gas Turbines
CCP	Climate Change Programme (UK)
CDU	Christian Democratic Union Party (Germany)
CEGB	Central Electricity Generating Board
CFCs	Chlorofluorocarbons
CH_4	Methane
CHP	Combined Heat and Power
Cispel	Confederazione Italiana Servizi Pubblici Enti Locali (Italy)
CO	Carbon monoxide
CO_2	Carbon dioxide
COGEN Europe	The European Association for the Promotion of Cogeneration
COP 1	Conference of the Parties 1 (Berlin 1995)
COP 2	Conference of the Parties 2 (Geneva 1996)
COP 3	Conference of the Parties 3 (Tokyo 1997)
DATAR	Délégation à l'aménagement du territoire et à l'action régionale (France)
DG	Directorate General (EU Commission)
DH	District Heating
DNC	Declared Net Capacity
DSM	Demand Side Management
DTp	Department of Transport (UK)
EAP	Environmental Action Programme (EU)
ECOFIN	Council of Economic and Finance Ministers (EU)
EdF	Electricité de France
EEO	Energy Efficiency Office (UK)
EFL	Energy Feed Law (Germany)
EFTA	European Free Trade Association
ENEA	National Agency for New Technologies, Energy and the

	Environment (Italy)
ENEL	Ente Nazionale Energia Elettrica (Italy)
ESEMA	Strategy for Energy and Environment (Spain)
EST	Energy Saving Trust (UK)
EU	European Union
FCCC	Framework Convention on Climate Change
FDP	Liberal Party (Germany)
FF	French Franc
FFL	Fossil Fuel Levy (UK)
FGD	Flue Gas Desulphurization
FNC	First National Communication (Italy)
FoE	Friends of the Earth
FRG	Federal Republic of Germany
GDP	Gross Domestic Product
GDR	German Democratic Republic
GWh	Gigawatt hour
GWP	Global Warming Potential
HCFCs	Halogenated Chlorofluorocarbons
HEES	Home Energy Efficiency Scheme (UK)
HFCs	Hydrofluorocarbons
ICE	Catalan Institute for Energy (Spain)
ICEP	Interministerial Committee on Economic Planning (Italy)
ICLEI	International Council for Local Environmental Initiatives
IDAE	Institute for Energy Diversification and Savings (Spain)
IEA	International Energy Agency
IEM	Internal Energy Market
IGC	Intergovernmental Conference
IGCC	Integrated Gasification Combined Cycles
IMA	Interministerial Working Group (Germany)
IPCC	Inter-Governmental Panel on Climate Change
IRP	Integrated Resource Planning
ISES	International Solar Energy Society
ktoe	Thousand tonnes of oil equivalent
kWh	Kilowatt hour
MCCC	Making a Corporate Commitment Campaign (UK)
MINER	Ministry of Industry and Energy (Spain)
MOPTMA	Ministry of Public Works, Transport and the Environment (Spain)
Mt	Million tonnes
MtC	Million tonnes of carbon
Mtce	Million tonnes of coal equivalent
Mtoe	Million tonnes of oil equivalent
MW	Megawatt
MW_e	Megawatts of electricity
MW_{th}	Megawatts of thermal energy
N_2O	Nitrous Oxide
NEP	National Energy Plan (Italy)
NFFO	Non-Fossil Fuel Obligation (UK)
NGOs	Non Governmental Organizations

NO_2	Nitrogen dioxide
NO_x	Nitrogen oxides
NUTEK	Swedish National Board for Industrial and Technical Development
O_3	Ozone
OECD	Organization for Economic Cooperation and Development
OFFER	Office for Electricity Regulation (UK)
OFGAS	Office of Gas Regulation (UK)
PAEE	Savings and Energy Efficiency Plan (Spain)
PEN	National Energy Plan (Spain)
PEN 91	National Energy Plan 1991–2000 (Spain)
PFBC	Pulverized Fluidized Bed Combustion
PGOU	General Urban Plan (Spain)
PNC	National Climate Programme (Spain)
PV	Photovoltaic
PWR	Pressurized Water Reactor
QMV	Qualified Majority Voting
RCEP	Royal Commission on Environmental Pollution (UK)
R&D	Research and Development
RECs	Regional Electricity Companies (UK)
SAVE	Specific Actions for Vigorous Energy Efficiency
SEA	Single European Act
SEK	Swedish Crown
SF_6	Sulphurhexafluoride
SEMAV	Secretary of State for Environment and Housing (Spain)
SNCC	Swedish National Committee on Climate Change
SO_2	Sulphur dioxide
SPD	Social Democratic Party (Germany)
SRO	Scottish Renewables Order (UK)
Swedish EPA	Swedish Environmental Protection Agency
TEU	Treaty on European Union
TGV	*Train à grande vitesse* (high speed train)
THERMIE	EU energy technology support programme
toe	Tonnes of oil equivalent
TPA	Third Party Access
TPES	Total Primary Energy Supply
TWh	Terrawatt hour
UBA	Umweltbundesamt (Federal Environmental Office) (Germany)
UNCED	United Nations Conference on Environment and Development
VAT	Value Added Tax
VDEW	Association of electricity companies (Germany)
VEAB	Växjö Energy Company (Sweden)
VOCs	Volatile Organic Compounds
VROM	Netherlands Ministry of Housing, Physical Planning and Environment
WMO	World Meteorological Organization

Chapter 1

THE CLIMATE CHANGE CHALLENGE

Ute Collier and Ragnar E Löfstedt

INTRODUCTION

The threat of large scale changes to the earth's climate over the 21st century has become one of the most salient issues in environmental policy during the 1990s. The main culprit is the fast increasing level of carbon dioxide (CO_2) emissions resulting from fossil fuel use. Energy production and use based on fossil fuels is an essential constituent of economic activities in the industrialized world and there are no easy recipes for emission reductions. Reduction strategies will have to be based on both technological and behavioural changes, with modifications to energy, transport and general economic policies. Climate change thus presents the clearest challenge yet for the integration of environmental concerns into other policy areas, as is advocated by the concept of sustainable development (see below). At the same time, the existence of so-called 'no regrets' policies (those which have no costs and sometimes even benefits for economy or society) has been confirmed by a number of studies, many of which point to feasible opportunities for achieving both environmental and economic objectives, for example through the vigorous pursuit of energy efficiency.

Despite great uncertainty about the exact scale and effects of climate change and the obvious challenge this issue poses, policy makers have committed themselves to action by adopting the Framework Convention on Climate Change (FCCC). This requires countries to draw up emission reduction programmes and currently negotiations are under way to find common agreement between signatory nations on specific emission reduction targets by 1997. The onus for emission reductions lies primarily on the industrialized countries, which are responsible for the greater part of past emissions. Currently, the US, Eastern Europe (including ex-USSR) and Western Europe together account for over half of global emissions (World Resources Institute, 1996), with per capita emissions particularly high in the US, where rates are nearly five times the global average. In the future, it is expected that China and India will quickly advance to the major emitter league, although it will take some time for their per capita emissions to equal those of the developed world. While the US is by far the largest emitter (25 per cent of CO_2 emissions), the 15 Member States of the European Union (EU) together are also responsible for a large proportion of emissions (almost 15 per cent), with per capita emissions well above the global average. Furthermore, the EU presents the only supranational level of policy making at which legal enforcement is possible, thus, at least in principle, ensuring the fulfilment of the environmental objectives agreed. As such, the setting of emission reduction targets and strategies at the EU level can also play an important role in setting a global example and brokering international agreements.

The EU has officially recognized its global responsibility in this area and already by 1990 had set itself a target to stabilize CO_2 emissions by the year 2000, thereby stressing its global leadership role. However, progress with this target and the implementation of a joint strategy has been limited (see Chapter 4). Effectively, the example of the EU epitomizes the problems of finding a global solution to the climate change problem. Per capita emissions vary considerably between countries, as do the opportunities for emission reductions. Southern European countries (in particular Spain and Greece) have pushed for exemptions as they bear much less responsibility for past emissions than do Northern European countries, especially the large emitters, Germany and the UK. Furthermore, there are also some EU specific problems in that, while environmental policy increasingly has acquired an EU dimension over the past 20 years, energy and transport policies have essentially remained in the national domain. Member States, particularly France and the UK, are jealously guarding their national sovereignty in these areas and have thus hampered progress in the climate change area. Furthermore, in recent years, in particular in view of the subsidiarity debate, doubts have been raised about the appropriateness of EU level action in this and other environmental policy areas.

The heterogeneous nature of emission characteristics and the potential for abatement, indicate that climate protection has to be a multilevel policy, with a supranational framework set at the global and EU level and sufficient flexibility for action at national, regional and local levels. Furthermore, the pervasive nature of emissions requires action in a variety of areas to minimize policy tensions and employ a broad range of policy instruments. These requirements imply the need for policy coordination and policy coherence between different levels of policy making, as well as between different policy areas.

The aim of this book is to examine climate change policy making in the EU in some detail, to identify complexities, constraints and opportunities, as well as to highlight the interaction between technological, economic and political factors. The main emphasis is on the presentation of detailed case studies of six EU Member States, which together account for more than three quarters of CO_2 emissions in the EU. Both national and local level activities are examined, with particular strengths and weaknesses being identified. Conclusions are drawn as to what can be learnt from diverse national and local experiences, and what role the EU might play in the future development of climate change strategies. This volume should be of interest to those working specifically in the area of climate change policy, as well as those concerned with the study of European integration and the interaction of policy making at different levels of governance. As such, this volume has a dual purpose as a study of responses to the climate change issue and as a case study of EU environmental policy making.

The remainder of this introductory chapter covers the essential background to the climate change issue, examining some of the available scientific evidence, the international response to the problem and the specific challenges presented to EU policy making in this area.

THE THREAT OF CLIMATE CHANGE

There are a number of so-called greenhouse gases in the earth's atmosphere which influence its climate. The most important ones are carbon dioxide (CO_2), methane (CH_4) and nitrous oxide (N_2O), but chlorofluorocarbons (CFCs) and halogenated CFCs (HCFCs) also play a role. The radiative forcing (the way in which these gases influence atmospheric processes) and persistence in the atmosphere of these gases vary, so scientists have developed indices to ensure comparability. According to these indices, CO_2 is the most important gas, accounting for 64 per cent of the total radiative forcing of all greenhouse gases, followed by CH_4 (19 per cent), N_2O (5.7 per cent) and then a number of other greenhouse gases (IPCC, 1996a). However, there are substantial uncertainties associated with the measurement greenhouse gases other than CO_2, in particular CH_4.

Globally, 87 per cent of the CO_2 emissions originate from industrial activities, in particular fossil fuel combustion, with the remaining 13 per cent due to deforestation (World Resources Institute, 1996). In the industrialized world, over 90 per cent of CO_2 emissions are due to energy production and use, with the main culprits being electricity generation, industrial and domestic energy use and transport related energy consumption (OECD, 1994). The reduction of CO_2 emissions has thus been the focus of policy responses to the issue, which is reflected in the CO_2 focus of this book. A proportion of CH_4 emissions is also energy related (leakages from coal mining and gas pipelines, flaring of natural gas at oil fields), with the rest being from various agricultural activities (rice paddies, enteric fermentation by livestock). N_2O emissions result mainly from various industrial processes.

Measurements show that the atmospheric concentrations of CO_2 and the other trace gases have undergone large increases since pre-industrial times. This has led to an enhancement of the naturally occurring greenhouse effect (whereby gases in the atmosphere prevent a proportion of the heat from the sun being re-radiated from the planet back into space), warming the surface and producing other changes of climate (IPCC, 1996a). These changes are expected to intensify and could potentially have serious consequences for humanity and the political stability of the world as low lying countries flood, cultivation patterns shift and large areas suffer from desertification. However, because of the complexity of atmospheric science, with a multitude of possible feedback mechanisms, there is much uncertainty as to the exact nature of these changes.

Most of the world's eminent climate scientists are collaborating under the auspices of the Inter-Governmental Panel on Climate Change (IPCC), which was set up in 1988 by the United Nations to provide scientific and policy advice in preparation for a climate convention. Experts are nominated by governments to participate in the IPCC's three working groups, to produce a synthesis of the state of knowledge in the following areas:

■ Working Group I: The science of climate change.
■ Working Group II: Impacts, adaptation and mitigation.
■ Working Group III: Economic and social dimensions.

Since the late 1980s, substantial effort and financial resources have been spent on climate change research and modelling, both in the natural and the social sciences, and considerable progress has been made in the understanding of climate change. Many of the initial projections for temperature and sea level rises have been scaled down but they are still expected to be serious enough to warrant abatement policies. In its second assessment report in 1995, the IPCC for the first time made a cautious link between the climate extremes of recent years and theories about climate change. While stressing that there are still many uncertainties, the summary of the second report for policy makers claims that 'the balance of evidence suggests a discernible human influence on global climate' (IPCC, 1996a).

Although based on a consensus, this finding has since been contested. It has been suggested that the final version of the *IPCC 1995* report lacks many of the references to uncertainties made in the draft report and thus skews the assessment towards sounding much more definite about the occurrence of climate change (Energy Economist, 1996). However, Brack and Grubb (1996) argue that the final version still remains opaque about the degree of confidence to be attached to human attribution. Furthermore, both government representatives and key scientists are united in support of the version published. In general, it must be stressed that all of the IPCC agree with the IPCC conclusions. Those who have been most vocal in their criticism have been discredited through the large sums of 'research money' obtained from coal and oil companies (Ozone Action, 1996).

Based on current emission trends, the IPCC 'best estimate' models project an increase in global mean surface air temperature of about 2°C by 2100, which would be an average rate of warming greater than any seen in the past 10,000 years. As a result of the accompanying thermal expansion of the oceans and melting of glaciers and ice sheets, sea levels are expected to rise by around 50 centimetres in the same time period. Other projected features include increased precipitation and soil moisture in high latitudes during winter, a greater frequency of extremely hot days and more extreme rainfall events (IPCC, 1996a).

Natural ecological systems, socioeconomic systems and human health are all sensitive to both the magnitude and rate of climate change. However, little is known about the exact nature of such sensitivities, thus making the prediction of impacts just as difficult as the prediction of effects. Shifts in vegetation zones, regionally differentiated changes in crop productivity, flooding in some areas and increased desertification are just some of the possible impacts mentioned by the IPCC (1996b), all of which have ecological and economic implications.

To summarize, the climate change issue is characterized by a high level of scientific uncertainty, as well as by potentially costly and even life threatening consequences. A response to the problem can be justified on account of both the preventive and precautionary principles, the main principles that guide EU environmental policy.

OPTIONS FOR EMISSION REDUCTIONS

The all-pervasive effect of CO_2 emissions presents some challenges for emission abatement, as there are no simple 'end-of-pipe' solutions. In the EU, electricity generation is one of the main CO_2 sources, accounting for around 31 per cent of total CO_2 emissions (Eurostat, 1996). CO_2 removal from power station emissions is technically possible but extremely costly, incurs severe conversion efficiency penalties and causes insurmountable problems in terms of disposal. Nevertheless, large reductions can be achieved through improving generating efficiencies, employing more cogeneration and switching fuels. Within the fossil fuel generation technologies, gas fired plants have some clear advantages over coal fired plants because of the lower carbon content of natural gas compared to that of coal and also because of higher generation efficiencies in modern combined cycle plants. Even greater emission reductions are possible by a switch to carbon-free generating sources (carbon free only applies to electricity generation itself – lifecycle emissions through building plants, fuel transport, etc can be considerable), namely nuclear power or renewable energies. Nuclear power continues to be one of the most controversial energy sources but apart from political acceptability problems, doubts have, in recent years, increased about its economic viability. Economic viability has also been a major factor influencing the development of renewable energy sources, although here the widely differing potentials between different countries and regions are also an issue.

Domestic energy use has stabilized over the past decade but, together with the tertiary sector, the domestic sector still accounts for 20 per cent of emissions in the EU (Eurostat, 1996). Around 60 per cent of this energy use is related to heating (OECD, 1994) and hence crucially dependent on insulation levels. Sometimes, independent of climatic variations, some countries have much higher insulation standards than others, for example Denmark compared to the UK. As is discussed in more detail in Chapter 2, there is much potential for efficiency improvements in buildings, as well as in water heating, lighting and electrical appliances.

Emissions from the industrial sector have declined in importance, to some extent because of efficiency improvements, but mainly because of fundamental changes in industrial structure, as most EU economies are moving from a manufacturing sector economic base to a service sector economy. That said, industry still accounts for 23 per cent of CO_2 emissions (Eurostat, 1996) and energy intensive sectors such as chemicals, steel, pulp and paper and the cement industry play an important role in a number of countries, in particular Germany, the Netherlands and Sweden. Although there has been much technological improvement over the past two decades, large energy efficiency potentials are still believed to exist, for example of 40 per cent in the cement industry or 20–30 per cent in the steel industry (OECD, 1994). The possibilities for emission reductions from energy activities are examined in more detail in Chapter 2.

Emissions from the transport sector are of particular concern, currently accounting for 26 per cent of CO_2 emissions in the EU. Growth in this sector, in total volumes, distances travelled and emissions (not only CO_2 but also various air pollutants), has been extremely fast, with no sign of abating. In the EU, transport emissions increased by an average of 3 per cent per annum between 1974 and 1992. Of all modes, air transport has seen the largest growth, almost doubling its share of CO_2 emissions to 2.6 per cent of the total (European Commission, 1994). Furthermore, nitrogen oxides (NO_x) emissions placed in the upper troposphere by aircraft are producing ozone (O_3), which at this atmospheric level acts as a greenhouse gas. It has been suggested that this could lead to a global warming impact as great as that from CO_2 emissions. The effect of aircraft on cloud formation also has a global warming impact (OECD, 1994).

Scenarios of future transport activity indicate that transport energy use in the industrialized nations could rise by between 40 per cent and 100 per cent by 2025, mainly in road based transport (IPCC, 1996b). Undoubtedly, both the private motor car and road freight transport have an inherent attractiveness because of the great flexibility they offer. According to the IPCC (1996b), improved vehicle energy efficiency might reduce greenhouse gas emissions per unit of transport activity by 20–50 per cent in 2025 relative to 1990. If users were prepared to accept changes in vehicle size and performance, transport energy intensity could be reduced by 60–80 per cent in 2025. However, it is not clear how these higher figures could be achieved realistically and there is an emerging consensus as to the need for some fundamental changes to modal shifts and the overall demand for travel and transportation. Land use planning has an important role to play in this context, as is further explored in Chapter 3.

DEVELOPING RESPONSE STRATEGIES

While technically there are thus various possibilities that would reduce CO_2 emissions, there are a number of economic and political constraints that impede a shift to these technologies. Assuming that the threat of climate change is serious enough to warrant a policy response, policy makers face a number of challenges. Most significantly, it is important to recognize that abatement cannot be confined to environmental policy and the application of isolated environmental policy instruments. As emissions are almost exclusively due to energy and transport activities, they are fundamentally shaped both by the operation of the market and by government policies in these areas. For example, support mechanisms for domestic coal in Germany and Spain have resulted in a high proportion of electricity being generated by coal, with correspondingly high CO_2 emissions (see Chapters 5 and 9). Furthermore, until the early 1990s, an EU directive was in force that banned the use of natural gas in electricity generation. In the transport area, various government policies favouring the private car at the expense of public transport have contributed to making this sector the fastest growing source of CO_2 emissions.

Responses to the climate change issue, therefore, crucially depend on emission reduction objectives being integrated into other policy areas, especially energy and transport. This requires both procedural and institutional changes, so as to ensure policy coordination and coherence (Collier, 1994). Here the climate change issue links into the wider framework of sustainable development, which, broadly speaking, aims to reconcile environmental protection with economic development. The sustainable development issue reached the political agenda at the same time as the climate change issue, with the adoption of Agenda 21 at the Rio Summit. Agenda 21 aims to provide a basis for a new global partnership for sustainable development and environmental protection and its signature commits countries to the drawing up of national sustainable development programmes (UNCED, 1992). While sustainable development itself is a rather vague concept and subject to a variety of interpretations, it is nevertheless useful in that it is providing an impetus for environmental objectives to be considered in different policy areas.

Sustainability has both ecological and economic dimensions, but in the case of energy and transport activities there can be substantial overlap between environmental protection and economic development objectives. In the energy area, large levels of inefficiencies, on both the production (especially electricity generation) and end use sides make neither environmental nor economic sense. At the same time, the vast increases in transport volumes across Europe have economic, as well as environmental, implications. For example, in the UK the economic losses due to road congestion have been estimated at £15 billion (Royal Commission on Environmental Pollution, 1994). Other impacts such as noise and accidents also have social costs (Maddison et al, 1996). Overall, the IPCC has estimated that greenhouse gas emissions in most countries can be reduced by 10–30 per cent at negative or zero costs, the so-called 'no regrets' measures (IPCC, 1996c). Furthermore, the aggregate net damage likely to be caused by climate change provides an economic rationale for going beyond 'no regrets' measures.

While many studies have shown the potential compatibility between each of climate change policies, other environmental protection aims and economic objectives, the development of response strategies nevertheless faces a number of obstacles. Firstly, there is the issue of external costs and other market failures. Environmental and social costs are generally inadequately reflected in energy pricing and as a consequence the market system presents no incentives to integrate these costs into decision making. The application of environmental taxes to the use of certain energy sources (carbon taxes in the case of measures to reduce CO_2 emissions) is aimed at dealing with this problem. Suggestions have also been made that there are opportunities to simultaneously meet climate change and employment generation objectives through a broad reform of the tax system, with the imposition of higher tariffs on energy use and the reduction of the tax burden on labour for both employer and employee. This has been described as a 'double dividend' (Carraro et al, 1995; DRI, 1994). However, as is discussed in subsequent chapters, the application of such taxes has proved difficult, both at EU level and in some Member States, partially on account of their distributive effects. Additionally, there are a number of other market failures which result in a suboptimal allocation of resources, for example as far as energy efficiency is concerned. This issue is explored in more detail in Chapter 2.

Secondly, there is a time dimension which, linked to the high level of uncertainty inherent in the issue, has implications for policy making. Despite the acceptance of the preventive and precautionary principles as the basis of environmental policy (for example in the Treaty on European Union), policy makers remain prone to short term thinking, which has been exacerbated by the market orientation of so much policy making in recent years, in particular in the UK (Hutton, 1996) and other countries with a strong emphasis on neoliberalism. A major issue is that of short term costs versus long term economic benefits, which exists for a number of abatement options. Within this context, individual countries are particularly concerned about the potential threat to their economic competitiveness, an issue which is heavily pushed by the industrial lobby. A commitment to joint action, ideally at the international level, but also within smaller supranational fora such as the EU, is crucial to solving this dilemma.

EMERGING POLICY RESPONSES: THE INTERNATIONAL LEVEL

Despite the large uncertainties in this important issue, and in view of the potentially disastrous consequences, concern emerged in some countries during the late 1980s that precautionary and preventive action is necessary. It is difficult to pinpoint conclusively the key event that brought the issue onto the international political agenda but expanding scientific evidence together with a major drought that hit the US in 1988 appear to have been instrumental (Rowlands, 1995). Obviously, the prospect of climate change is a global environmental problem, so that action at this level is crucial for solving the problem. Negotiations for a global convention started in 1990 and, within an almost record time in international negotiation terms, the FCCC was signed by 165 nations at the Rio Summit in 1992, 154 of which have since ratified it, including the 15 EU Member States. According to the convention, the aim should be to achieve:

> *...the stabilization of greenhouse gas concentrations in the atmosphere at a level that would prevent dangerous anthropogenic interference with the climate system. Such a level should be achieved within a timeframe sufficient to allow ecosystems to adapt naturally to climate change, to ensure that food production is not threatened and to enable economic development to proceed in a sustainable manner (United Nations, 1992).*

In terms of commitments, the drafting of the FCCC is somewhat vague, requiring industrialized countries to draw up abatement programmes which should aim at 'returning individually or jointly to their 1990 levels those anthropogenic emissions of CO_2 and other greenhouse gases not controlled by the Montreal Protocol'. No specific timeframe for achieving this 'stabilization' is included, although there is a mention elsewhere in the convention that stabilization by the end of the 1990s is desirable (for more detail on the convention see Pitt and Nilsson, 1994; Rowlands, 1995; Rowbotham, 1996).

By the time of the First Conference of the Parties (COP 1) in Berlin in 1995, it became clear from the national programmes which had been submitted that most signatories would not achieve this stabilization. Furthermore, the IPCC had always stressed the need for real emission reductions. Before COP 1, calls for a protocol to specify emission reductions became ever stronger and the Alliance of Small Island States (AOSIS), supported by the environmental lobby, proposed a protocol calling on industrialized countries to reduce their 1990 level of CO_2 emissions by at least 20 per cent by the year 2005. The protocol found a surprising level of support but in the end faltered in the view of opposition from the oil-producing nations as well as the US.

Nevertheless, there was a general acceptance for the need for emission reductions, and the compromise outcome of the meeting was the Berlin Mandate, which specified that a binding protocol should be developed by 1997, to contain 'quantified limitation and reduction objectives within specified timescales'. An ad hoc group was established to prepare the protocol and its first meetings showed continuing differences in opinion between the negotiating parties. At the time of writing, a successful outcome was thus far from certain.

In July 1996, the Second Conference of the Parties (COP 2) took place. This conference occurred in the aftermath of the second IPCC report (specifically Working Group 1) which argued that evidence for climate change is stronger than ever before. It was hoped in some circles that such evidence would lead countries to commit themselves to large CO_2 cuts post 2000. However, although the world's largest emitter, the US, announced that cuts post 2000 were necessary and should be legally binding, no agreement was reached as to their actual level, and it was decided that any proposed cuts should be put forward at COP 3, which will be held in Tokyo in 1997. A report of the meeting in *The Economist* was scathing, stating that while discussions are going on as to how best to reduce emissions beyond the year 2000, little effort was made to stabilize emissions at 1990 levels by the year 2000, as suggested in the FCCC. Instead, since 1990 emissions have increased by 4 per cent in the Organization for Economic Cooperation and Development (OECD) (*The Economist*, 1996). If nations are indeed to be taken seriously in their attempt to reduce CO_2 emissions post 2000, stabilization by 2000 is the minimum that must be achieved.

In recent years, there has been an increasing recognition that across-the-board emission targets are difficult to achieve, even within the industrialized countries, in view of differences in starting points, approaches, economic structures and resource bases, as recognized in the Berlin Mandate [Article 4.2(a)]. An alternative would be differentiated emission targets, although the negotiation of such targets would in itself be politically difficult. It thus seems likely that, if a protocol with specified reduction objectives is agreed, it would at best differentiate between industrialized and developing countries. However, as the EU is a signatory to the Convention, as well as all its Member States, there is scope for target differentiation, and even emission trading, within the EU context.

THE EU RESPONSE

Differences in starting points have made an agreement on common targets and actions in the EU difficult to reach. There is an EU target for stabilization by 2000 (based on 1990 levels) but this is based on individual countries' commitments (or lack of in some countries), rather than a specific agreement on target sharing. Subsequent efforts at drawing up a strategy with common measures for emission reductions have hit a number of obstacles, in particular the lack of EU competences in the fields of energy and transport policy. Furthermore, the centrepiece of the proposed strategy was a carbon/energy tax which in some ways was doomed to failure considering past problems with attempts at fiscal harmonization. Effectively, the EU climate change strategy consists of some guidelines on energy efficiency measures, targets for renewable energies and support programmes for the development of energy technologies. None of these are expected to make a substantial impact on emissions, as Chapter 4 demonstrates.

The climate change issue has coincided with a growing dissatisfaction in some Member States with the role of the EU and the centralization of policy making. As far as environmental policy is concerned, this dissatisfaction has different causes. Some of the Northern, more environmentally proactive Member States have become disillusioned with the lowest common denominator approach of many EU environmental measures. Conversely, the Southern Member States have been concerned about the costs imposed on them by EU environmental legislation. Furthermore, the UK has become increasingly vocal against EU involvement in general (Collier and Golub, 1997). As a result, the issue of subsidiarity, referring to action at the lowest possible level, has increasingly influenced policy making. In the case of climate change, this has had a crucial impact on the fate of the carbon/energy tax, as well as on action with regard to energy efficiency, as Chapter 4 discusses in more detail.

Subsidiarity is not necessarily a bad principle to apply in climate change policy making, considering the differing starting points and reduction potentials found in the 15 Member States. However, as Collier (1996) has argued, the subsidarity discussions have been driven primarily by political and economic expediencies rather than a real concern for other issues such as environmental efficiency. Nevertheless, the issue of the appropriate role for EU level intervention is an important one and is addressed in the conclusions (Chapter 11).

One issue of concern, which is seldom discussed, is that the process of European integration has in itself negative implications for climate change because of the implicated large increases in production, trade and associated levels of transportation. According to Gabel (1996), trade liberalization within the EU is expected to increase the demand for road transport by 34 per cent. As mentioned before, emission reductions from this sector have to address the issue of overall demand and it is not easy to see how these two issues can be made compatible. However, while such broader issues need to be addressed seriously, there is certainly scope for action in a variety of areas, not all of which result in such policy tensions. Obviously, the EU Member States are more likely to initially employ a 'no regrets' strategy.

NATIONAL AND LOCAL STRATEGIES

The 15 Member States certainly vary in a number of important areas, including commitment to the issue, past efforts in relevant fields (such as energy efficiency), fuel shares in energy consumption and levels of economic development. In terms of overall CO_2 emissions, Germany emerges as the worst offender, followed by the UK, Italy, France and Spain. The five largest Member States thus account for three quarters of total CO_2 emissions, although in terms of per capita emissions, the smallest member state, Luxembourg, is in the lead, followed by Denmark. The most important determinants of these differences are as follows:

- fuel shares in electricity generation;
- level of economic development;
- industrial structure and industrial energy efficiencies;
- energy consumption for heating;
- car ownership and modal splits in transport;
- overall distances travelled.

Considering the heterogeneous emission characteristics displayed by Member States, there is limited scope for harmonized policy approaches. One of the main roles of the EU in recent years has been to monitor national programmes through the monitoring mechanism established in 1992. This requires Member States to annually submit information on emissions and progress with their national climate change strategies. According to the latest report under the mechanism (European Commission, 1996), there are substantial comparability problems in terms of emission trajectories, as well as insufficient information about the implementation of measures. As a result, the Commission considers it is impossible to evaluate the effectiveness of individual countries' policies.

There is a need for independent assessment of individual countries' climate change strategies to allow for a more effective monitoring of compliance with the FCCC. In fact, this book has grown out of a project for the European Commission that was to provide detailed case studies of a number of EU countries. Considering the size of the EU, a trade-off has been made between providing detailed information on a selected number of Member States or a more cursory treatment of all Member States. It was decided to focus this study on a sample, thus allowing a much more detailed assessment of policy developments. For such a sample, the above main emitters are clearly most interesting. However, it was also decided to include Sweden in the sample, as a country which is known to be particularly proactive on environmental issues but also because commitments to phasing out nuclear power and past efforts in energy efficiency mean that it faces a difficult task in further reducing CO_2 emissions.

Apart from looking at the national programmes, an analysis of climate change strategies also has to pay attention to the local level. Environmentalists have long argued that an increased local dimension is needed for environmental policies, as well as greater participation. This call has gained renewed currency with Agenda 21, which stresses the importance of the local level (UNCED, 1992). Local authorities in the EU have considerable influence over a number of activities which are important for CO_2 emissions, especially in the energy and transport fields. The scope for action depends very much on the exact nature of local authority competences, which vary from country to country. As far as energy is concerned, the potential influence is obviously greatest where there are municipally owned energy companies (eg in Germany, Austria, Sweden) which can directly influence investments and pricing and thus have an impact on emissions, provided the right priorities are being set (Collier and Löfstedt, 1996).

However, while there are some environmental and logistical arguments for policy action and planning at the local level, it is obvious that this cannot take place within a vacuum and to the exclusion of activities at other levels. Only a few reasons are mentioned here:

■ Energy demand is affected by overall macroeconomic policies.
■ Energy prices and hence incentives for energy efficient behaviour are partially a result of market forces, partially of central and/or regional government regulation.
■ Regional variations in renewable resources mean that some areas might find themselves totally reliant on other areas.

Higher level policy intervention is thus necessary to ensure fairness and because economic and political realities may impede effective action at the local level. While local government may be strong and promoting environmental protection in some Member States, such as Germany, in others it has fewer powers, fewer resources and/or the environment is not a priority. There are also certain potential policy instruments (such as efficiency standards or carbon taxes) which have to be developed centrally and cannot be drawn up separately by each local authority. Finally, in many countries local authorities have no influence over energy companies which are regulated by national (or sometimes regional) governments. Climate change policy thus has to be a multilevel policy, with coordinated action at all levels, applying a variety of policy instruments. In reality, as this book shows, policy coordination and coherence is currently not being achieved in the EU. Current approaches also lack commitment and fail to seriously pursue the wider issue of sustainable development.

THE FOLLOWING CHAPTERS

The aim of this chapter is to introduce the main issues of relevance to the analysis of climate change policies in the EU. Many of the issues touched upon are examined in more detail in subsequent chapters. Chapter 2 deals with the energy dimension of climate change policies, looking at the potential for emission reductions through various technological and other options. Chapter 3 provides the same analysis for transportation issues. This is followed in Chapter 4 by an examination of the EU's attempt to develop a common response to the climate change issue. Chapters 5–10 examine the climate change strategies of six EU Member States in more detail. The order of the country case studies is in relation to their overall importance in emission terms: Germany followed by the UK, Italy, France, Spain and Sweden. Finally, Chapter 11 focuses on a comparative analysis and provides some final conclusions.

REFERENCES

Brack, D and Grubb, M (1996) *Climate Change: A Summary of the Second Assessment Report of the IPCC*, Briefing Paper No 32, July, Royal Institute of International Affairs, London

Carraro, C, Galeotti, M and Gallo, M (1995) *Environmental Taxation and Unemployment: Some Evidence on the Double-Dividend Hypothesis in Europe*, Nota di Lavora 34.95, Fondazione Eni Enrico Mattei, Milano

Collier, U (1994) *Energy and Environment in the European Union: The Challenge of Integration*, Avebury, Aldershot

Collier, U (1996) 'Developing responses to the climate change issue: the role of subsidiarity and shared responsibility' in Collier, U, Golub, J and Kreher, A (eds) *Subsidiarity and Shared Responsibility: New Challenges for EU Environmental Policy*, Nomos Verlag, Baden/Baden

Collier, U and Golub, J (1997) 'Environmental policy and politics' in Rhodes, M, Heywood, P and Wright, V (eds) *Developments in West European Politics*, Macmillan Press, London

Collier, U and Löfstedt, R (1996) 'Think globally, act locally? Local climate change strategies in Sweden and the UK', *Global Environmental Change* Vol 6, no 4, forthcoming

DRI (1994) *Potential Benefits of Integration of Environmental and Economic Policies*, Graham and Trotman, London

Energy Economist (1996) 'Climate change 1995', *Energy Economist*, Briefings Supplement, May 1996

European Commission (1994) '1993 – Annual energy review', *Energy in Europe*, special issue, June

European Commission (1996) 'Second evaluation of national programmes under the monitoring mechanism of Community CO_2 and other greenhouse gas emissions', *COM* (96), 91

Eurostat (1996) *Annual Energy Review 1995*, Office for Official Publications, Luxembourg

Gabel, L (1996) 'The environmental effects of trade liberalization: a review of some studies', paper presented at the Deregulation and the Environment workshop, European University Institute, 9–11th May 1996

Hutton, W (1996) *The State We're In*, Cape, London

IPCC (1996a) *Climate Change 1995: The Science of Climate Change*, Cambridge University Press, Cambridge

IPCC (1996b) *Climate Change 1995: Impact, Adaptions and Mitigation of Climate Change: Scientific and Technical Analyses*, Cambridge University Press, Cambridge.

IPCC (1996c) *Climate Change 1995: Economic and Social Dimensions of Climate Change*, Cambridge University Press, Cambridge

Maddison, D, Pearce, D, Johansson, O, Calthrop, E, Litman, T and Verhoef, E (1996) *Blueprint 5: The True Costs of Road Transport*, Earthscan, London

OECD (1994) IEA/OECD *Scoping Study: Energy and Environmental Technologies to Respond to Global Climate Change Concerns*, OECD, Paris

Ozone Action (1996) 'Scientific skeptics on attack at international meeting on climate change', *Ozone Action Media Release*, 10th July 1996

Pitt, D and Nilsson, S (1994) *Protecting the Atmosphere: The Climate Change Convention and its Context*, Earthscan, London

Rowlands, I (1995) *The Politics of Global Climate Change*, Manchester University Press, Manchester

Rowbotham, E (1996) 'Legal obligations in the climate change convention' in O'Riordan, T and Jäger, J (eds) *Politics of Climate Change: A European Perspective*, Routledge, London, pp32–50

Royal Commission on Environmental Pollution (1994) *Transport and the Environment*, HMSO, London

The Economist (1996) 'Tempestuous', *The Economist*, 20th July, p86

UNCED (1992) *The Global Partnership for Environment and Development: A Guide to Agenda 21*, UNCED, Geneva

United Nations (1992) *Framework Convention on Climate Change*, United Nations, Geneva

World Resources Institute (1996) *World Resources 1996–1997*, Oxford University Press, Oxford

Chapter 2 | **THE ENERGY DIMENSION**

Ute Collier and Ragnar E Löfstedt

INTRODUCTION

Over the past decade, the environmental impacts of energy production and use have become increasingly apparent. While there are solutions to many of the problems associated with the industry, these are often complex and involve high costs. This is particularly true for measures conceived in response to the climate change issue. Addressing this issue first and foremost requires reductions in fossil fuel use, on which the EU energy sector is highly dependent. In the EU as a whole, 95 per cent of CO_2 emissions are due to fossil fuel combustion, of which around 25 per cent are related to the transport sector, the remainder resulting from various energy related activities, as Figure 2.1 shows.

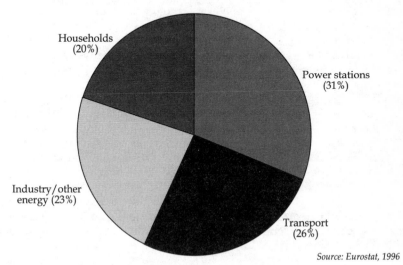

Source: Eurostat, 1996

Figure 2.1: EU CO_2 emissions by sector (1994)

The contribution of the energy sector to CO_2 emissions varies between the EU Member States. For example, in Sweden and France, where most electricity is produced from hydro or nuclear power, CO_2 emissions are substantially lower on a per capita basis than are those in Germany, the UK (see Chapters 5 and 6) or Denmark, where power generation is based on fossil fuels.

This chapter gives an up-to-date assessment of the scope for reducing CO_2 emissions within the EU energy sector, based on EU specific economic and political factors, as opposed to general technical potentials which have been the focus of many other studies [eg International Energy Agency (IEA), 1991; IEA, 1994a, Krause and Koomey, 1994a; Pearman, 1992). The roles of CO_2-free electricity generating sources, such as hydro and nuclear power, and the new and developing renewables are examined. The importance of energy efficiency and the impacts of switching from fossil fuels that emit high levels of CO_2, such as coal, to those with low CO_2 emissions, such as natural gas, are also discussed. Most of these options also bring with them other environmental benefits, such as reduced air pollution. Furthermore, there are economic and employment benefits, especially as far as energy efficiency measures are concerned, and as such CO_2 abatement is reconcilable with a general move towards a more sustainable energy system. Generally, the CO_2 issue can be seen as providing a welcome stimulus for change in an inefficient and environmentally damaging energy system. However, some technologies which are beneficial in CO_2 terms involve other environmental impacts, which need to be taken into account in policy making.

A particular emphasis of this chapter is on the options available to reduce emissions from electricity generation, as opposed to those related to heat production. There are several reasons for this emphasis:

■ It is on the whole considerably easier to replace heat produced from fossil fuels (often with electricity itself) than electricity generated from fossil fuels.
■ Together with transport, electricity generation is the fastest growing source of CO_2 emissions in Europe, expected to increase by 7.9 per cent between 1992 and 2000 (European Commission, 1995).
■ In final energy demand, the prevailing trend is for a growing penetration of electricity, as it has become increasingly important in industrial processes.
■ There is also growing demand in the domestic and tertiary sector as the move to an 'information society' involves a rapidly growing use of computers, while air conditioning is becoming increasingly popular, not only in Southern Europe.

Meanwhile, the demand for heat has not increased. Also, not all European nations need large amounts of heat (some nation's climates are warmer than others), but all nations need a certain amount of electricity.

THE ROLE OF CO_2-FREE ELECTRICITY GENERATING SOURCES

Currently, 47 per cent of the EU's electricity is generated from non-fossil fuel sources, almost exclusively in large hydro and nuclear power plants. The increase in nuclear generating capacity over the past 20 years or so has decreased the CO_2 intensity of each unit of electricity generated, a trend which is particularly marked in France (see Chapter 8). In principle, there is considerable potential to increase the role of these sources. However, in practice there are major obstacles to their further development: they are not commercially viable for one reason or another, or already fully developed or unpopular among the general public.

Nuclear Power

Nuclear power plants constitute around a quarter of the EU's overall electricity generation capacity, but the future of this generating source is uncertain. The only country in Europe which is still pursuing a nuclear expansion programme is France. Currently, France has 56 nuclear reactors in operation with a net capacity of 58.5 GW, with four further reactors under construction which by 1998 will add 11.2 GW (see Chapter 8). There are several reasons for the French enthusiasm for nuclear power. The French economy was severely affected by the 1973 oil embargo as the country was highly dependent on imported oil for electricity production at the time, as it has no indigenous fossil fuel reserves. To increase its energy independence, France, in a similar fashion to Sweden, embarked on an intensive nuclear power expansion programme (Jasper, 1990), so that by 1995 81 per cent of the nation's electricity production was generated from nuclear power, making it the highest proportion in the world after Lithuania (IEA, 1995). The nation's nuclear expansion policy has been helped by the fact that the utility that builds and operates the plants, Electricité de France (EdF), is a government owned monopoly, and has had no problem in accessing capital for its investments. The French public are also largely indifferent to the safety and waste problems posed by nuclear power, an issue that has not gone unnoticed in other parts of the EU or the US; this indifference has certainly been a factor in France's rapid nuclear power growth (Slovic, 1987; Slovic, 1993).

In the rest of the EU, nuclear power generation capacity is progressively disappearing, as it is now seen to be largely uneconomic when compared with cheap fossil fuel generating capacity (particularly based on North Sea and Russian natural gas). It also has a negative image in the eyes of the general public, which is not only concerned about the safety aspects of nuclear power plants and the problems associated with nuclear waste, but also believes, in some cases, that the regulatory agencies and the industry itself have not told the whole truth (Wynne, 1982). The unpopularity of nuclear power is demonstrated by the results of several national referenda. In Italy, following years of public concern, nuclear power was phased out in the late 1980s. In Sweden, following a referendum in 1980, all of the country's 12 nuclear power stations will be phased out by the year 2010 (see Chapter 10). Similarly, Spain has decided upon a moratorium for new nuclear plants. In Germany, public opposition to nuclear power has a long history and has been successful in halting the commissioning of a number of reactors (see Chapter 5). More recently, regulatory authorities have been finding it increasingly difficult to dispose of their nuclear waste in designated waste sites, as was recently exemplified by the public outcry and violent protests that surrounded nuclear waste disposal in Gorleben. In Finland in the early 1990s there were plans to build a fifth nuclear power station, but this was vetoed by parliament. Plans to build nuclear power stations in Denmark were shelved in response to public concern as early as 1978 (Jamison et al, 1990).

Interestingly, in the UK, where government ministers once promised 'electricity too cheap to meter' from nuclear power (Jasper, 1990), economic realities are proving to be the downfall of the nuclear industry. There have been two attempts to privatize the nuclear power industry, first in 1989 along with rest of the electricity industry and then in 1996 as a separate entity with a new name, British Energy. In the first case privatization failed due to the venture being largely unprofitable, despite a generous public subsidy system (see Chapter 6). With the second attempt, the Government decided effectively to 'sell out' the nuclear industry, the total price of privatization being less than the cost of building the nation's latest (and probably last) nuclear plant (Sizewell B) in order to appease city analysts who are concerned by the uncertainties posed by the unknown but potentially high decommissioning costs. The new company has already announced that any further investment in nuclear capacity is unlikely and it is instead proposing to build gas fired plants.

The contribution of nuclear power to future EU carbon abatement strategies, therefore, is unlikely to be substantial. The debate surrounding climate change has had little effect on the fortunes of nuclear power and it is a link that few nuclear proponents have made. Furthermore, several studies (Jackson, 1992; Krause and Koomey, 1994b) have claimed that nuclear power, when compared with improved energy efficiency, is not a cost effective CO_2 abatement option. Overall nuclear power generation capacity in Europe is set to decline in the medium term, as virtually no nuclear power stations are being built and the nuclear plants currently in operation are retired due to old age or shut down due to public opposition. In fact, the running down of the EU's nuclear power capacity will have negative consequences for its CO_2 abatement strategy because it is increasingly being replaced by fossil fuel capacity. An illustrative example of the effect of this policy is given by Sweden, where plans are well advanced to phase out one, possibly two, nuclear plants over the next two years and replace them with either electricity imported from Norway or electricity generated in Sweden using natural gas. As a result, CO_2 emissions will increase in either Norway or Sweden. A similar scenario will occur in the UK if nuclear plants are gradually phased out and replaced by natural gas plants (see Chapter 6). This exemplifies the dilemma nuclear power poses. On the one hand it is rejected by many on environmental grounds because of the inherent risks and potential disastrous consequences of an accident. On the other hand, in CO_2 terms, a phase out is not desirable.

Hydropower

Hydropower, the EU's second most important CO_2-free electricity generating source contributes around 13 per cent of total electricity generation. It also faces an uncertain future. Most of the commercially viable hydropower sites in the EU have already been developed with those remaining being uneconomic in current market conditions (European Commission DG XVII, 1994). Power companies today, operating in an increasingly competitive climate, and at a time of low fossil fuel prices, are unable to justify the large capital required to build new large head hydropower stations. City analysts want short term, high profit return on any capital outlays, skewing generating capacity to natural gas powered combined cycle turbines (CCGTs) (see p24).

Most of the remaining hydropower potential is found along the large, free flowing rivers in northern Scandinavia which are protected for cultural, environmental and recreational (eg salmon fishing) purposes (Vedung, 1984). This is not to say that the future of hydropower in Europe is static. Although there has been some concern about the environmental effects of expanding hydropower in Norway, energy policy makers in that country, as well as in Iceland, maintain that there is still a huge untapped hydropower capacity. Over the past five years there has been considerable discussion about selling surplus hydropower to the continent via electricity cables. Furthermore, elsewhere there is still potential for small and medium sized hydropower plants, which are also less capital intensive.

THE POSSIBLE CONTRIBUTIONS OF NEW AND DEVELOPING RENEWABLE ENERGIES

Considering the lack of scope for expansion in nuclear and hydropower, reducing CO_2 emissions in the EU energy sector will have to rely on other options. For this, the new and developing renewables offer some promise, and have long been the favourite energy option of environmentalists. The renewables considered here include those which currently show most promise: solar, biomass, wind and geothermal sources.[1] In the EU in 1992 such sources accounted for just under 6 per cent of primary energy production, with 95 per cent of this deriving from biomass (European Commission, 1995). Austria, Denmark, Finland and Sweden, nations that are either richly endowed with forests, peat or wind, are the leaders in developing renewables in Europe. These same nations have tried to make renewables more

---

1 Wave and tidal power could also be promising but still require a large amount of research and development (R&D) before they can become candidates. In many studies, energy from waste is also included in this category. However, considering that much domestic and industrial waste consists of materials made from nonrenewable resources (such as plastic), energy from waste does not as such qualify as a renewable energy.

commercially attractive through environmental taxes on fossil fuels. In Sweden, for instance, CO_2, sulphur dioxide (SO_2) and NO_x taxes on fossil fuels mean that biomass generated heat can compete commercially with coal and oil in the district heating sector. This has resulted in a virtual doubling of the use of wood fuels from 5.4 to 9.8 TWh over a two year period from 1993 to 1994 (see Chapter 10).

Clearly, the technical potential for renewables in the EU is considerable. Estimations vary from study to study but a wide ranging assessment for the Commission completed in 1994 suggests a technically accessible resource of up to 47 per cent of 1990 final energy consumption (European Commission DG XVII, 1994), although this includes large scale hydro power sources. The exploitation of such a large resource is subject to various constraints, mainly economic[2] but also, paradoxically, environmental. Recent years have seen great improvements in the renewables policies of many countries, as the case study chapters show more clearly. Currently, the new and developing renewable energies are still heavily dependent on government support programmes, although in some cases (especially wind power), there is a slow move from R&D programmes to actual market introduction policies.

Some analysts are very optimistic about current developments. In a recent far reaching study on renewable energy sources, Grubb (1995) argues that renewables have a bright future in Europe. Total EU R&D spending on them has virtually trebled, and there is a new found interest in renewables even now at a time of low energy prices. Grubb suggests a number of reasons for this. Countries in the EU have a series of environmental targets that they have to meet, and diversifying the energy mix will make it easier to reach these targets. Additionally, the EU is looking for ways to reduce its agricultural subsidies. Taking land out of production and encouraging the planting of energy forests and various energy crops will reduce the (expensive) agricultural surplus at the same time as increasing renewables in the energy equation. Finally, investments in renewables create jobs. Recent studies show that for every 1 MW of installed biomass capacity, three jobs are created (Knutsson, 1995).

There are some clear success stories. In Austria, for example, the Government decided to encourage the use of biomass for heating purposes on a micro scale to reduce the residues from the country's timber industry. A result of this programme was that by 1990 solid biomass supplied virtually a quarter of Austria's domestic heating needs (2.5 Mtoe) and this is set to double by the year 2005 (Grubb, 1995).

Germany has seen a real 'wind boom' with an increase in installed capacity of 78 per cent between 1994 and 1995 to 1127 MW.[3] However, as Chapter 5 discusses, this development, apart from state subsidies based on favourable rates for access to the grid, has been challenged by the powerful energy company lobby. Also, despite the expansion, the new and developing renewables still account for less than 0.5 per cent of German primary

---

2 For example, the cost of electricity from PV power is currently 8–10 times that of electricity from conventional plant.

3 *Renewable Energy Report*, 24th May 1996, No 15, p 6

energy production (IEA, 1995). A similar pattern of events is occurring in Spain. The Spanish R&D programme is focusing on small-head hydropower, solar photovoltaic (PV), biomass and wind energy and aims to increase the contribution from renewables from 0.8 Mtoe to 1.9 Mtoe by the year 2000. However, this is insignificant when one considers that during the same period it is projected that natural gas use will increase from 5.7 Mtoe to 13.5 Mtoe, and that CO_2 emissions will increase by at least 25 per cent by the year 2000 based on 1990 levels (see Chapter 9; IEA, 1995).

Furthermore, other environmental factors are becoming increasingly problematic. In the UK, for example, many wind power projects that received government approval under the Non-Fossil Fuel Obligation (NFFO), have not received planning permission because of public opposition on account of factors such as visual intrusion and noise (see Chapter 6). Siting problems are also becoming increasingly apparent in Germany and Denmark.

The main conclusions of this section are thus relatively pessimistic, especially for the short to medium term. A study for the European Commission has projected a renewables contribution of only 6.4 per cent of primary energy production under existing programmes by 2010 (European Commission DG XVII, 1994). With other measures (such as a carbon/energy tax), a contribution of 9.2 per cent could be achieved, but many assumptions of this scenario are not being met currently. Another Commission scenario is considerably more optimistic, predicting a contribution of 13.9 per cent by 2010 (European Commission, 1995). However, this is based on an assumed reduction in overall primary energy production and a large growth in imports, which skews the picture. In terms of final energy demand, only 4.9 per cent would be met by renewables, even under the most optimistic scenario.

In general, one cannot forget that previous European and North American expectations have been overly optimistic.[4] In reality, despite some successes, in current market conditions renewables are not able to compete with fossil fuels in the heating or electricity sectors. Lower energy prices, due to the increasingly competitive nature of the energy market, are presenting yet another obstacle to renewable energies. There is an urgent need for a better internalization of the external costs of other forms of energy production, be it in terms of energy taxes or through 'externality adders' where there is a procurement system for new capacity. Meanwhile, the amounts invested so far are minimal compared to other R&D and subsidy programmes (eg fusion research and coal subsidies). As a result, with the exception of the Scandinavian countries and possibly Austria, new and developing renewable energy sources will, in present market conditions, help little in reducing the EU's future CO_2 emissions.

---

4 An example of this is the former US president Jimmy Carter's renewable programme which aimed to have 20 per cent of the nation's energy mix derived from renewable sources by the year 2000. Based on current trends, the likely figure will be 10 per cent (Grubb, 1995).

ENERGY EFFICIENCY

The above discussion clearly suggests that the scope for expansion of CO_2-free generating sources to enable emission abatement is limited. A more promising area is the possibility of emission reductions from fossil fuel sources by minimizing energy demand through improvements in the efficiency of energy use. Efficiency improvements allow environmental, security of supply and resource conservation objectives to be met simultaneously and are a prerequisite for a renewables based system, which is unlikely to meet current levels of demand. Many energy efficiency investments also make economic sense, although less so in periods of low energy prices, such as being experienced currently.

In the industrial sector, the past 20 years have seen substantial improvements in the energy efficiency of production processes. Between 1973 and 1988 (in general a period of high energy prices), the energy intensity of the manufacturing sector in Europe decreased by 34 per cent (IEA, 1994a). However, as there has been a replacement of primary fuels by electricity in many production processes, CO_2 emissions have not decreased by the same amount but have to some extent shifted to the electricity generation sector. Nevertheless, the IEA estimates that there is a high potential for CO_2 reduction through fundamental process changes, as well as through materials recycling and substitution. These potentials vary substantially between countries and between sectors on account of different starting points and processes. For example, for the Netherlands, Worrel et al (1994) estimate a total technical potential for energy efficiency improvements of 31 per cent of primary energy consumption, with sectoral variations between 8 per cent (nonferrous metals) and 45 per cent (food and beverage). An additional saving of 5 per cent could be made through the wider application of combined heat and power (CHP) (further discussed on pp24–25). A greater use of recycled secondary materials and feedstock can also yield emission reductions (IPCC, 1996).

In the area of domestic energy efficiency there is a clear North–South divide. Sweden, Denmark, Finland and Germany all have high building standards in place, making homes and offices energy efficient (in Sweden, for instance, three pane windows are standard), while in Southern Europe, in countries such as Spain and Italy, and in the UK building standards are lax. There are several reasons for these differences. In Scandinavia and Germany high building standards have reduced dependency on oil use for heating, thereby securing some energy independence, while in the warmer climate of Southern Europe heating bills are not such an economic burden and the risk of ill health from poorly heated homes is low. Consequently, the need for strict building standards is not perceived as a priority. In the UK, the low standards are partially due to the more temperate climate (at least compared to Scandinavia) but mainly because the domestic fossil fuel supply means that fuel prices are comparatively low, thereby limiting the incentive to save energy.

The potential for further energy efficiency in Europe is considerable. It has been estimated that in the UK 30 per cent of domestic household energy could be saved simply through implementing Scandinavian type building standards (Nakicenovic, 1989). In Sweden, which already has strict energy efficiency standards in place, 15 per cent savings in electricity use could be achieved through technical and economic measures particularly in the appliance and electricity sectors (Ling and Wilhite, 1990; Mills, 1990). The IEA (1994a) has estimated the following potential energy savings:

- 30–70 per cent household appliances;
- 10–50 per cent domestic water heating;
- 70–90 per cent existing buildings.

There are clear technical potentials for energy efficiency improvement, but realizing them is impeded by a number of obstacles, mainly related to market failures. The IEA (1991b) has identified the following market barriers:

- lack of knowledge;
- separation of expenditure and benefit;
- limited capital;
- rapid payback requirements;
- energy tariff structures;
- lack of interest;
- legal and administrative obstacles.

These obstacles have, in the main, been recognized by governments and addressed in their energy policy through a variety of measures such as grants and subsidies, information campaigns and tariff restructuring. However, in many countries funding for energy efficiency programmes has been cut in recent years, as later chapters show. A greater involvement of energy companies in promoting energy efficiency, akin to the US model of integrated resource planning with demand side management (DSM), has been suggested as a way forward in this area (Collier, 1994; Woolf and Mickle, 1993). A proposal to this end (further discussed in Chapter 4), was made by the European Commission in 1995 but it currently seems unlikely to make much progress, due to opposition from the powerful energy lobby. As Boyle (1996) shows, experience with DSM has been gained in a number of EU countries, although it has been somewhat piecemeal. This generally shows that there is a need for supportive policy measures, a clear motivation for the utility and/or pressure from consumers or advocacy groups.

Overall, the technical potential for energy efficiency within the EU is considerable, potentially resulting in substantial CO_2 emission reductions. However, it will not be that easy to realize this potential due to the series of obstacles discussed above. These need to be addressed through a variety of measures, including incentive structures for energy companies and standards for a wide range of appliances. Also, the issue of low energy prices needs to be addressed. Chapter 4 examines the effort to impose a carbon/energy tax at the EU level. In order for this to occur, however, there has to be the political will, which, so far has been noticeably absent.

Nevertheless, there are some signs of a revival of the energy efficiency theme in energy policy, mainly on account of the climate change issue; some CO_2 emission reductions will result from these efforts.

Low CO_2 Fossil Fuel Technologies

Meanwhile, currently the only area in which significant emission reductions are occurring is in electricity generation, mainly as a result of switching fuel from coal to gas and of the use of more efficient plant. Nowhere is this trend more pronounced than in the UK, where the large scale substitution of coal for natural gas in electricity generation is in process (the so-called 'dash for gas'). By the year 2000, more than 30 per cent of UK electricity will be generated from natural gas (see Chapter 6). The technological mainstay of this development is the CCGTs; these have CO_2 emission advantages both because of the lower carbon content of natural gas (0.61 tC/Mtoe) compared to coal (1.09 tC/Mtoe for bituminous coals; Grubb, 1990), and the higher efficiencies compared to a normal coal plant.[5] As a result, new CCGTs can reduce carbon emissions by more than 60 per cent compared to existing coal plants (Krause and Koomey, 1994a). Furthermore, these plants are also economically attractive as they can be built in smaller modules (thus particularly suitable for investors in a competitive energy market), have short lead and construction times, and offer a cost effective option to comply with air pollution legislation.

CO_2 emission levels can be lowered even further by gas fired CHP plants. CHP (although mainly based on coal) has long been established in a number of EU Member States. In Denmark, Finland and the Netherlands up to 35 per cent of electricity is produced in CHP plants, which are also well established in Austria and Germany. In these countries, most CHP plants are associated with district heating grids, but some also have industrial applications. CHP plants produce heat in addition to electricity, thus exploiting the fuel more fully than do electricity-only plants, where more than 60 per cent of the fuel (in conventional coal plants) is dissipated in the form of low grade heat. Even with existing coal fired CHP systems, a 44 per cent reduction in CO_2 emissions can be achieved compared to a conventional coal plant (Krause and Koomey, 1994a).[6]

Currently, CHP generation probably accounts for little more than 7 per cent of electricity produced in the EU (COGEN, 1995), although the figures are somewhat unreliable due to a lack of proper accounting. If CHP appears so attractive for environmental and resource conservation reasons, the question arises why it is not more widespread. COGEN Europe, the European Association for the Promotion of Cogeneration, in a recent study identified a number of obstacles, including lack of political will, high infrastructure costs

---

5 The best new coal-fired plants achieve net efficiencies of 39.5%,while CCGTs currently operate at efficiencies of 52–55%, with one manufacturer, Asea Brown Boveri (ABB), even offering a CCGT reported to achieve 58% efficiency (Krause and Koomey, 1994a).
6 Figures for industrial cogeneration coal turbines suggest CO_2 emissions of 159 gC/kWh compared to 280 gC/kWh for an existing coal steam turbine.

for district heating systems, unfavourable regulatory systems, low tariffs for cogenerated electricity and lack of third party access to the national grids (COGEN, 1995). Several countries have instigated programmes and policies for CHP, although often only partially removing these obstacles. The case studies in subsequent chapters of this book discuss CHP developments in individual countries in more detail. For the EU as a whole the outlook for CHP, in particular industrial cogeneration, is positive, with CHP likely to account for 11 per cent of new generating capacity installed between 1996 and 2000 (European Commission, 1995); it also stands to benefit from a more liberalized energy sector. Nevertheless, further policy measures need to be implemented to remove the remaining obstacles.

There are also a number of advanced coal technologies (for example PFBC – pulverized fluidized bed combustion, or IGCC – integrated gasification combined cycles) which offer CO_2 advantages over conventional coal plants, although at a much lower level than those of gas plants. With current low gas prices and the short construction times for CCGT modules, the latter are far more economically attractive. However, the 'dash for gas' has not been universally welcomed. Resource conservation and energy security of supply considerations largely have been ignored. Fells and Woolhouse (1994) point out that this reflects a rather short-term perspective as North Sea natural gas reserves will be depleted within 50 years.

New Opportunities in a Liberalized Energy Market?

The European energy market is currently in a state of flux. Over the past six years there have been various attempts to introduce competition as well as privatization into the energy market. So far, the results of this have been very variable across the EU. The UK energy sector is now completely privatized, although competition will not be fully in place until 1998 (see Chapter 6), while in Sweden deregulation of the electricity market and competition was introduced in January 1996, leading to foreign firms purchasing large parts of Sweden's generating sector (see Chapter 10). Similar developments have been seen in Norway and Finland, with privatization and deregulation likely to follow in several other EU states, most notably Italy and Spain (see Chapters 7 and 9).

Liberalization has, by some, been hailed as a great environmental opportunity (eg Flavin and Lenssen, 1995). However, liberalization can have both positive and negative implications for climate change. On the one hand, it promises a great opportunity to shake up a monopolistic energy market that is dominated by supply oriented companies and focused on large scale technologies. This could provide a chance for renewables and DSM. On the other hand, a 'free for all' situation could mean damaging competition between options such as gas versus district heating or CCGTs versus renewables. Grubb (1995) argues that in the short term financial perspective, liberalization runs counter to the strategic and externality benefits that form a major justification for renewable energies. The best outcome in environmental terms can thus not be assumed, especially while multiple market failures exist, including those related to external costs.

An attempt to examine some of the consequences of liberalization in the electricity sector has been made by the IEA (1994b). According to this study, the effects of liberalizing energy markets tend to include:

- less investment in large generating plant;
- increased diversity of investors, technology and fuels;
- competition between fuels in the heating market;
- increased development of smaller units, typically closer to load centres.

Investment decisions in a liberalized energy market will tend towards technologies that are less capital intensive and that have shorter lead times. Furthermore, increasing the number of investors can also encourage the consideration of less traditional, financially riskier technologies. The evidence of the effects of liberalization to date is somewhat mixed. While the 'dash for gas' has occurred in the UK, in Sweden there have been attempts to market certain types of electricity as more environmentally friendly (offering consumers green electricity generated by hydropower rather than nuclear or fossil fuels). This latter move is interesting as it is seen to increase environmental awareness among the general public, possibly leading to changes in energy consumption habits in the long term (see Chapter 10).

A potential problem lies with the key aim of liberalization to reduce energy prices as a means of increasing the competitiveness of European industry. Low energy prices have been widely recognized as a main obstacle to greater energy efficiency, so unless other incentives can be created liberalization is likely to have a negative effect as a strategy for reducing CO_2 emissions. Nevertheless, a framework for DSM could be created with competitive bidding procedures as used in the US (Collier, 1994).

As has been pointed out already, it is likely that privatization and liberalization will lead to short term CO_2 reductions via fuel switching from coal to natural gas. In the medium to long term the likely train of events is unclear due to uncertainty about the general energy policy climate and levels of environmental concern. It is feasible that, in some countries, security of supply concerns, which have recently become somewhat unfashionable, might return to the agenda and hence affect fuel choices. In the UK, the Labour front bench energy spokesman, John Battle, has made it clear that his party would like to see a more balanced energy strategy in the UK and is keen to support a move to clean coal technology (Löfstedt, 1996).

CONCLUSIONS: WHAT DOES THE FUTURE HOLD?

Overall, despite some positive indications in a number of areas, there seems to be limited potential for significant reductions in CO_2 emissions within the EU energy sector in the current policy climate. With nuclear power gradually being phased out in many EU countries for one reason or another, the expansion of hydropower at a standstill and a lack of political will to either promote energy efficiency programmes or invest substantially in renewable energy sources, it would be surprising if the EU was able to reduce, or indeed stabilize, CO_2 emissions in the medium term. Stabilizing CO_2 emissions by the year 2000 based on 1990 levels may be possible, but this would be mainly due to short-term gains resulting from Germany and the UK (Europe's two largest CO_2 emitters) switching from coal to natural gas. Switching to gas will eventually reach a resource constraint limit and even relatively efficient gas fired plants still emit considerable amounts of CO_2. The essential point is that gross domestic product (GDP) is expected to grow continuously and with it, even if at a lower rate, energy consumption and CO_2 emissions. A total decoupling of CO_2 emissions from economic growth does not currently seem feasible.

Recent EU Commission projections based on a conventional wisdom scenario support these arguments. According to this model, energy related CO_2 emissions are set to increase by 14.1 per cent between 1992 and 2020. While the growth in the transport sector will be fastest (22.3 per cent), energy sector emissions are nevertheless set to grow by 12.6 per cent, with the growth related to electricity generation equalling that of the transport sector, despite a doubling of the contribution of natural gas (European Commission, 1995). Such projections, despite the obvious uncertainties associated with them, throw serious doubts on the feasibility of post 2000 emission reductions. The Commission forecasting exercise also included one scenario in which an 11 per cent reduction in CO_2 emissions is achieved by 2020. This 'Forum' scenario is based on assumptions about a strong role for public administration and intervention, including vigorous energy efficiency policies, taxation and a climate conducive to energy investment (eg low discount rates). Its assumptions about nuclear power (a capacity increase of 188 per cent in the time period) are clearly subject to substantial criticism. However, the general message is one of policy intervention and behavioural changes, issues which need to be kept in mind when assessing current policies in the Member States.

REFERENCES

Boyle, S (1996) 'DSM progress and lessons in the global context', *Energy Policy*, Vol 24, no 4, pp345–359

COGEN (1995) *The Barriers to Combined Heat and Power in Europe*, COGEN Europe, Brussels

Collier, U (1994) *Energy and the Environment in the European Union*, Avebury, Aldershot

European Commission DG XVII (1994) *The European Renewable Energy Study (Main Report, Annex 1, Technology Profiles)*, European Commission, Brussels

European Commission (1995) 'European energy to 2020', *Energy in Europe*, special issue, December 1995

Eurostat (1996) 'CO_2 emissions from fossil fuels 1985 to 1994', as quoted in *European Environment*, 30th April 1996, ppI.1–3

Fells, I and Woolhouse, L (1994) 'A response to the UK national programme for CO_2 emissions' *Energy Policy* Vol 22, no 8, pp666–684

Flavin, C and Lenssen, N (1995) *Power Surge: A Guide to the Coming Energy Revolution*, Earthscan, London

Grubb, M (1990) *Energy Policies and the Greenhouse Effect, Volume One, Policy Appraisal*, Royal Institute of International Affairs, London

Grubb, M (1995) *Renewable Energy Strategies for Europe: Volume 1*, Earthscan, London

IEA (1991a) *Greenhouse Gas Emissions: The Energy Dimension*, OECD, Paris

IEA (1991b) *Energy Efficiency and the Environment*, OECD, Paris

IEA (1994a) IEA/OECD *Scoping Study: Energy and Environmental Technologies to Respond to Global Climate Change Concerns*, OECD, Paris

IEA (1994b) *Electricity Supply Industry: Structure, Ownership and Regulation in OECD Countries*, OECD, Paris

IEA (1995) *Energy Policies in IEA Member States*, OECD, Paris

IPCC (1996) *Climate Change 1995: Impact, Adaptions and Mitigation of Climate Change: Scientific and Technical Analyses*, Cambridge University Press, Cambridge

Jackson, T (1992) *Efficiency without Tears: 'No-Regrets' Energy Policy to Combat Climate Change*, Friends of the Earth, London

Jamison, A, Eyerman R, and Cramer J (1990) *The Making of the New Environmental Consciousness*, Edinburgh University Press, Edinburgh

Jasper, J M (1990) *Nuclear Politics: Energy and the State in the United States, Sweden and France*, Princeton University Press, Princeton

Knutsson, G (1995) 'Lägesrapport beträffandet programmet för energieffektivisering och ökad användning av förnybara energiresurser för ett miljöanpassat energisystem i Osteuropa', Discussion Paper from NUTEK's Baltic/Eastern Europe Programme, Stockholm

Krause, F and Koomey, J (1994a) *Fossil Generation: The Cost and Potential of Conventional and Low-Carbon Electricity Options in Western Europe*, International Project for Sustainable Energy Paths, El Cerrito

Krause, F and Koomey, J (1994b) *Nuclear Power: The Cost and Potential of Conventional and Low-Carbon Electricity Options in Western Europe*, International Project for Sustainable Energy Paths, El Cerrito

Ling, R and Wilhite, H (1990) 'Norwegian electrical appliance ownership, family types, and potential energy savings' in *Proceedings of the American Council for an Energy Efficient Economy–Human Dimensions*, American Council for an Energy Efficient Economy, Washington DC

Löfstedt, R E (1996) 'Blir Battle Blairs nye energiminister?' *Electricitetens Rationella Anvaendning* Vol 4, pp32

Mills, E (1990) *An End-Use Perspective on Electricity Price Responsiveness*, Vattenfall, Vällingby, Sweden

Nakicenovic, N (1989) *Technological Progress, Structural Change, and Efficient Energy Use – Final Report*, International Institute for Applied Systems Analysis, Laxenburg

Pearman, G I (1992) *Limiting Greenhouse Effects; Controlling Carbon Dioxide*, John Wiley & Sons, Chichester

Slovic, P (1987) 'Perceived risk' *Science* Vol 236, pp280–285

Slovic, P (1993) 'Perceived risk, trust, and democracy' *Risk Analysis* Vol 13, no 6, pp675–682

Vedung, E (1984) 'Striden om de strommande vattnen' In Tekniska Museet (ed) *Daedalus, Tekniska Museets arsbok 1984*, Tekniska Museet, Stockholm

Woolf, T and Mickle, C (1993) *Integrated Resource Planning: Making Electricity Efficiency Work*, Greenpeace International, Amsterdam

Worrell, E, de Beer, J and Blok, K (1994) 'Abatement of CO_2 emissions by energy efficiency improvement', *Industry and Environment* Vol 17, no 1, pp32–35

Wynne, B (1982) *Rationality and Ritual: The Windscale Inquiry and Nuclear Decisions in Britain*, British Society for the History of Science, Monograph 3, Chalfont St Giles, Bucks

Chapter 3 | # ROUTES TO REDUCING TRANSPORT-RELATED GREENHOUSE GAS EMISSIONS

Jo Smith and Susan Owens

Introduction

Transport, and specifically road transport, has been cited for over 30 years as the most challenging of environmental problems. The same period has seen rapid growth in both private and freight road transport. So many aspects of our lives – from the development of the global economy to individual life chances – are shaped by access to a fossil fuel based transport system. Environmental imperatives have not proved sufficiently forceful to bring about a reassessment of this system. Current understanding of climate change may provide the most compelling case yet for a new approach to gaining access to basic needs and wants.

Transport is of particular significance to climate change policy makers because, in addition to being a major contributor to emissions of CO_2 [24 per cent in the UK (Department of the Environment, 1994b), 20–30 per cent in the other EU Member States (Hewett, 1995)], it is also, on current trends, the fastest growing sector in terms of these emissions (Department of the Environment, 1994b; Goodwin, 1994). Road transport is of particular significance, contributing 87 per cent of transport related emissions in the case of the UK (RCEP, 1994).

This chapter outlines the evolution of car dependence, pointing to how planning served the pressures for increased provision for road travel throughout the post World War II period. It outlines the basket of policy options that are currently being considered in EU countries. The main body of the chapter focuses on the particular contribution of land use planning to reduce the CO_2 emissions associated with transport. After a brief account of the traditional objectives of land use planning, the new theory and practice that seek to overturn previous planning priorities are considered.

THE EVOLUTION OF CAR DEPENDENCE

Throughout the developed world, the post World War II period has seen remarkable growth in both private and freight road transport. A cycle developed wherein increased demand for motorized mobility was reflected in policy priorities, notably road building programmes. At the same time individual decisions that shape land uses, relating to both domestic and industrial needs, were to some extent 'freed' from the constraints of distance.

Suburbanization and the development of smaller independent dormitory communities outside cities were thus catalysed by technological possibility, personal preferences and state regulation (planning and associated policy). The pattern is uneven across Europe, with some countries displaying higher densities in terms of built urban form and others with more dispersed rural populations. Nevertheless, the general trend towards excessive travel by car in cities and their immediate surroundings is true across the EU (ECMT and OECD, 1995). Together, these related forces have ensured that over the past 50 years, people's mobility choices have increasingly been *reduced to the choice to drive*. Trip patterns have, quite literally, been set in concrete.

The resulting physical separation of housing, employment, welfare needs, shops and so on has been furthered by the (related) tendency towards concentration of services into fewer larger units (Owens, 1995a). Transport models have tended to deal with broad distributions of population and employment, rather than the travel demand impacts of individual developments. There is, however, a developing body of data that shows development patterns to be an important influence on the mode of travel. In general, these patterns point to the persistence of car and lorry based transport planning (Newman and Kenworthy, 1989).

Private road transport has permitted the development of more complex trip patterns: for the car owning population, life choices are shaped by the car (Banister, 1991) and by policies that have been developed to serve it. This is reflected in the fact that in the South-East of the UK, typical journeys to work are 40 per cent longer than they were 20 years ago (SERPLAN, 1989). It is also important to note, however, that the importance of the journey to work has diminished, and now represents just 20 per cent of all trips (Banister, 1991). Leisure, other social activities and shopping trips have all become more frequent and longer, with the number of shopping trips doubling in the period 1965–1985 (Department of the Environment, 1988).

It is widely agreed that such trends in transport are not sustainable (Goodwin et al, 1991); however, it is also clear that the web of social, economic and political factors that have enabled the massive growth in motorized mobility is dense and resistant to change. Forecasts of road traffic continue, despite recession in European economies, to point to further increases in greenhouse gas emissions from the transport sector.

Sustainable Transport Strategies

The range of policy options available to policy makers is well rehearsed. There is a degree of consensus in the transport policy communities of EU countries as to the likely range of measures that might be drawn upon to respond to rising CO_2 emissions from the transport sector. These have tended to emphasize technological and traffic management systems. OECD countries have focused on:

■ improving the operational efficiency of transport systems;
■ increasing the energy efficiency of vehicles;
■ promoting of less carbon intensive transport fuels;
■ developing public transport and rail freight systems (Peake, 1996).

The 1995 IPCC assessment of options to reduce transport related greenhouse gas emissions puts much emphasis on the possibility for emission reductions through changes in vehicle intesity or energy sources. It suggested that (IPCC, 1996):

■ Improved vehicle energy efficiency might reduce greenhouse gas emissions per unit of transport activity by 20–50 per cent in 2025 relative to 1990 without changes in vehicle performance and size.
■ If users were prepared to accept changes in vehicle size and performance, transport energy intensity could be reduced by 60–80 per cent in 2025.
■ With energy intensity reductions, the use of alternative energy sources could in theory almost eliminate greenhouse gas emissions from the transport sector after 2025. A complete transition to zero greenhouse gas emission surface transport is conceivable, but would depend on eliminating emissions throughout the vehicle and fuel supply chain.

It is worth noting that there are some policy driven technological developments that are actually serving to increase CO_2 emissions from road transport. International success in noxious emission reduction and control has resulted in unanticipated and/or unintended second layer effects. Catalytic converters carry a modest fuel efficiency penalty. Regulation was in this case driven by public concern about local amenity and environmental issues. The easing of local air pollution problems may serve to reduce public pressure for action on traffic congestion. Similarly, some commentators have promoted diesel fuel as a more efficient alternative to petrol. However, recent research findings have raised public fears about the health impacts of urban diesel emissions. Such examples point to the complexity of transport, environmental and human health relationships, and the associated policy debates (Peake, 1996).

In addition to the above technological and infrastructural approaches, there is also a widespread recognition of the central importance of price signals in reducing CO_2 emissions related to transport. There are a range of approaches, but it is now common in EU countries to find that fossil fuel prices rise significantly faster than inflation, justified in terms of environmental policy.

There is a danger in presenting a summary of policy options as above: they might be viewed as a range of alternatives. It has, however, become a mantra of the environment transport debate that policies must be *integrated*, ie based on a strategic reassessment and shaping of technologies and a modal split by combining the range of market and regulatory measures. Hence pricing signals, improved public transport and integrated transport and land use planning aimed at reducing the need to travel must be considered mutually reinforcing (Owens, 1995a). In a major statement on both the nature of environmental problems arising out of transport patterns and their solution, the UK Royal Commission on Environmental Pollution (RCEP) stressed that its policy proposals: 'complement each other, and must be viewed as a whole' (RCEP, 1994).

Many of these policies aimed at adaptation and prevention of climate change promise far reaching social and economic implications, particularly for the transport sector. This is perhaps the best single explanation for the slow pace of agreement and implementation. However, it is relevant to note that climate change is only one of the reasons for controlling growth in number of vehicle miles travelled. It is not, however, clear that future policy will seek to reduce significantly the number of vehicle miles travelled. The goals of slowing and stabilizing road traffic growth, allied to increased efficiency measures, may serve to ameliorate some aspects of the 'transport crisis'. They will not in the near term deliver significant reductions in CO_2 emissions, which can only be achieved by means of significant reductions in road travel demand.

One of the most widely praised, but least tested, components of integrated transport policies is the integration of land use and transport planning with the purpose of reducing the need for motorized travel. It is praised for its simplicity and long term cost effectiveness. By shaping the development of land uses in favour of walking, cycling and public transport as the means of gaining access to needs and wants, planners can bring significant and lasting reductions in CO_2 emissions. Despite its potential, however, there are few examples of implementation. This is explained in part by the lack of a long term view of transport and land use policy, and also by broader countervailing social, political and economic trends.

Inertia, however, is not the sole reason for neglect of land use planning as a policy tool in strategies for reducing greenhouse gas emissions. Technology and market based solutions to (or corrections of) environmentally destructive impacts of the transport and energy sectors have been given much attention. The returns are near term, and the measures mirror tried and tested policy instruments. The impacts of environmentally oriented land use planning policies are almost impossibly difficult to anticipate or measure. The relations between transport, land use patterns and emissions are complex, and the factors that influence individual trip patterns and aggregate transport demand by mode are also difficult to research. While the long term rewards are great in terms of CO_2 emission reductions, short term costs can be significant. These considerations have stood in the way of both implementation and academic analysis. Recognizing this, the remainder of this chapter draws on case studies and the existing literature to give an account of the potential of land use planning, not as a 'lever' with its own share of anticipated impacts, but as one essential component among several that can deliver significant reductions in CO_2 emissions.

TRADITIONAL OBJECTIVES OF LAND USE PLANNING

Land use planning as policy and practice has long been driven by various purposes, paramount being social and economic goals. The former are expressed most strongly in the Garden City visions of Howard and Geddes (Hardy, 1991a). While these have proved influential, notably in the history of Swedish and British planning, the presumption to favour (economic) development has dominated.

The sustainability agenda undermines an assumption that has informed planning practice for decades: that environmental and social objectives must be 'balanced' by and in effect be subsidiary to the need to deliver the spatial conditions for economic growth. Indeed, planning is now expected to play a key role in the delivery of environmentally sustainable development, in terms of enabling sustainable use of water, minerals and energy resources (Rees and Williams, 1993; Owens, 1991) and nature and landscape protection (Countryside Commission, 1995).

Of more immediate relevance to a discussion of climate change are the claims made for planning as a means of reducing greenhouse gas emissions, primarily CO_2 (Owens and Cope, 1992; Countryside Commission, 1995). While there is a generally unmet potential in terms of CO_2 emission reductions relating to built form, the greatest potential reductions can be found in the transport sector.

The argument that land use planning can serve transport and energy demand reduction as effectively as it served the massive growth in demand in these sectors through most of the post World War II period is intuitively attractive. Indeed, these ideas have been translated into broad statements of intent, and some specific policies in several European countries including the Netherlands and the UK [Department of the Environment, 1994a, b; Netherlands Ministry of Housing, Physical Planning and Environment (hereafter referred to as VROM) 1990a, b; 1994a, b; see Smith and Owens, 1996, for a fuller account of the Netherlands example].

It is perhaps worth noting that the Netherlands' policy in this area, despite widespread recognition that it is more advanced than anywhere else in Europe, has not yet resulted in measurable reductions in total transport related CO_2 emissions. Throughout this discussion of the potential of land use planning as a policy tool in CO_2 emission reduction policies it is important to remain aware of the decisive importance of dominant trends in (global) political economy. The most pressing trend is pressure for increased economic growth and, by implication, trade and movement of goods and people. Land use planning tools will do no more than restrain travel demand growth until the assumption that economic growth implies transport growth is challenged.

Just as it has been an important factor in permitting private transport growth, land use planning has the potential to play a significant role in reducing the environmental impact of transport. Policies aimed at reducing the need to travel have been most visible at the level of the city. Examples include the Netherlands 'compact city' policy (ECMT and OECD, 1995), integrated land use and transport planning in the radial suburban developments of the Copenhagen 'Fingerplan' (City of Copenhagen, 1995) and in Stockholm and Vienna (ECMT and OECD, 1995), and control imposed on out of town retailing in the UK (Department of the Environment, 1994a), France and the Netherlands (TEST, 1989). It is important, however, to stress that such policies are being introduced against a backdrop of contradictory pressures that mean that they will only be able to stem the tide of traffic growth.

Nevertheless, for the purposes of this discussion it is worth looking at the estimates of fossil fuel energy savings (ie CO_2) that have been made. In theory, at least, estimates suggest that potential impacts, phrased as energy efficiency savings, compare well with those which might be achieved by more conventional measures. Transport energy demand varies by a factor of two or three between the most and least efficient land use patterns (Owens and Cope, 1992). The savings indicated are 'ideal': they would be expected to be less than this in practice as 'energy demand may not be as elastic as the models assume, the starting point may not be the worst case, and the optimum structure is unlikely to be achieved' (Owens and Cope, 1992).

LAND USE PLANNING AS A CLIMATE CHANGE POLICY INSTRUMENT

At the urban and regional scales, the single most important factor affecting travel needs is the physical separation of activities. The key variables in this relationship are density and the degree of mixing of different land uses (Owens and Cope, 1992).

While the policies implied by an integrated approach have implications for land use, indeed are reliant on their capacity to influence the balancing of locational and other factors, this discussion focuses on the capacity of land use planning itself to bring about changes in travel patterns. The planning system, including development control, might be framed in such a way as to encourage land use patterns that reduce travel demand and/or encourage more energy efficient modes.

Such a development of the role of land use planning represents a major shift away from structuring land uses in order to accommodate increases in road transport, and instead represents an attempt to curb it. This implies significant changes in planning education, as well as reappraisal of planning priorities at every scale. Experience suggests that legislation, or at least planning policy guidance, is a necessary component of this reordering of the purpose of land use planning. Although the theoretical assumption that energy use for transport falls as urban density increases is difficult to prove empirically, there is support for this from cross-sectional studies and analyses of travel patterns within large metropolitan areas (Newman and Kenworthy, 1989).

Increases in 'non motorized' accessibility can be achieved by bringing housing, employment and services together in relatively compact urban centres. A range of studies promote the concentration of development as a route to an energy efficient urban form (Owens, 1986); indeed the European Commission's Green Paper on the Urban Environment (Commission of the European Communities, 1990) endorses this view. There are counter arguments that focus on the potential negative consequences of 'town cramming'. This point, however, would be less relevant in a situation where the land take implications of large area private car road space and parking demands have been significantly reduced (Naess, 1991).

'Decentralized concentration' presents an alternative to the 'compact city' that also promises to bring increased energy and travel efficiency. This is achieved by decentralization of some jobs and services, and relating them to residential areas. This can be introduced either within a single large urban area or by creating independent settlements that may or may not remain linked to the original centre. These ideas have been promoted for over 20 years and there has been some implementation (see Owens and Cope, 1992, footnote 115).

The problems associated with this concept, however, are illustrative of the potential pitfalls of land use planning as a keystone of transport demand reduction. Much depends on individual responses to the restructuring of decision environments relating to travel. Trying to bring a reduction in travel demand in this way may incur social or economic costs for individuals or businesses in terms of amenity and services or access to jobs and markets. If travel costs do not carry sufficient deterrent, then decentralized concentration may actually result in less energy efficient land use and transport interactions, as both cross commuting and other travel increase to meet perceived needs (Owens and Cope, 1992).

The British case suggests that, in the absence of integrated transport policies, the latter has prevailed. Decentralization has, over the past few decades, resulted in rapid growth of suburbs and free standing settlements with the potential to be self contained. However, in the context of wider socioeconomic trends, the level of autonomy achieved has been limited, even where containment was a specific objective, such as in the 'new towns' (Hardy, 1991b).

The autonomy of settlements does tend to increase with size, but is also a function of relative isolation. In the South-East of the UK in Kent, for example, three quarters of the county's population live in 18 towns of over

15,000 people. While each town has a mix of facilities, their proximity has meant that 'the demand for movement between them is very high' (Kent County Council, 1990). At the same time, relatively remote new towns such as Newtown in mid Wales have achieved much higher degrees of autonomy (Breheny, 1990).

Clearly assumptions about energy efficient forms of urban development need to be carefully considered and placed in their socioeconomic context. Nevertheless, the evidence suggests that:

> *both centralization and 'decentralized concentration' are likely to be more energy efficient than peripheral or ex-urban residential development unrelated to jobs and services, but involve different sets of costs (Owens and Cope, 1992).*

Hence, in recognizing the need to provide both a sufficient range of job opportunities and services requiring only short journeys, as well as effective public transport networks, planners may look to development in existing centres rather than developing new settlements. Although there are promoters of new settlements as a means to energy efficiency gains (Blowers et al, 1993), there is clearly a case for saying that these gains will only be delivered where sites are in relatively remote areas.

Land use planning has the potential to play a significant role in encouraging the shift to more energy efficient forms of transport (RCEP, 1994). By catalysing reductions in distance to work and services, planning can encourage the 'green modes' of walking and cycling. For 'modal shift' policies to succeed, it is also necessary that the interests of pedestrians and cyclists be reflected by positive discrimination in favour of these modes in integrated land use planning, ie improvements in the cleanliness and safety of walking and cycling.

Modal shift from cars to public transport can also be facilitated by land use planning when it is integrated effectively with transport planning practice. By considering transport and land use implications of new developments together it is possible to promote those patterns that are better suited to the efficient and economic operation of public transport modes. Hence, whereas it is difficult to serve dispersed low density residential areas with public transport, relatively concentrated provision of housing and facilities permits the high loads that are essential to the environmental and economic efficiency of public transport (Owens and Cope, 1992). By discouraging dispersed low density residential areas some degree of concentration (though not necessarily centralization) that favours public transport can be achieved. Linear urban forms with moderate overall densities can be achieved by encouraging, for example, broad bands of urban development that combine high densities along public transport routes.

CONCLUSION: PLANNING FOR ACCESS AND REAL WEALTH

There is clear evidence that land use planning can play a major role in reducing CO_2 emissions relating to the transport sector, but it should be considered a necessary rather than sufficient component of CO_2 emission reduction strategies (Owens, 1995b). For land use policy instruments to be effective, the land use planning and transport system must be seen as a complex whole. Reducing the environmental impact of transport requires an integrated approach that combines planning, other forms of regulation, market mechanisms and public participation, as well as communication of the need to reduce road transport demand.

While the proclaimed consensus around a 'new realism' (Goodwin et al, 1991) on transport and environmental issues disguises important differences of opinion, notably on the feasibility of 'sustainable mobility' as promoted by the European Commission (Commission of the European Communities, 1992), it is clear that there is a broad based political and social recognition of the need for more integrated approaches to transport policy.

What is specifically important about the role of land use planning in this new approach to providing access to need is that it is politically less visible and contentious than market measures. It is also relatively inexpensive and highly effective over the longer term, and delivers a range of positive secondary impacts relating to quality of life. While it is impossible to meaningfully quantify the economic and social costs and benefits of an 'access' as opposed to 'mobility' based transport policy against a background of rising energy prices, land use policies aimed at reducing the need to travel could gather a great deal of commercial and public support.

Hence the key role that planning can play is in overcoming a 'prisoner's dilemma' that is a central obstacle to effective climate change policy. In the current circumstances rational self interest in terms of transport decisions does not produce the best outcome, even, in many ways, for the individual. To some degree no choice exists: in the current circumstances individuals and businesses are forced to spend significant time and money on road transport. Added to this, we have become deeply attached to the privacy and personal liberty afforded by private transport. Hence our preferences as citizens for social and environmental objectives can be satisfied only by policies that both act to contain our current preferences as consumers and to deliver real alternatives to motorized road transport.

If land use planning is to play its proper role, coordination between local, strategic and regional scales in land use, transport planning and other policy areas is essential. Of particular importance is the capacity to develop integrated strategies that encompass the wider city region. It follows from this that for such tools to be effective, their main mode of delivery – the local state – must be given legitimate authority to ensure that planning priorities are driven by environmental objectives. Hence, planners, politicians and other relevant decision makers must assess the land use, transport and energy implications of proposed economic developments according to global, regional and local environmental impacts.

It must be recognized, however, that the tendency to view economic and environmental interests as in conflict persists, despite universal support for the principles of 'sustainable development'. The rhetoric of sustainable development must now be translated into planning policy and practice that delivers increased environmental security and quality as a result of economic development. In this sense environmental priorities in planning can be portrayed not as limiting but rather as shaping choices about the manner and location of economic development.

Confidence persists in policy circles of the potential of technical fixes to permit continued high levels of motorized mobility. However, the long lead times that are required if land use policy is to have any impact upon CO_2 emissions, combined with the precautionary imperative that current understanding of climate change promotes, suggest that land use policies must be implemented with as much urgency as any proposal. Indeed, clear long term commitments to such policies at European, national and local scales will serve to shape technological research and development in ways that maximize the social, environmental and economic benefits of environmentally oriented land use planning strategies. These strategies would follow general guidelines for planning practice, including the following:

- Planning policies should encourage development in centres large enough to provide good access to a range of jobs and services without the need for regular private car journeys.
- There should be a presumption against dispersed low density residential areas and against service and employment developments that depend on car use.
- Developments expected to generate a large amount of traffic should be integrated with cycle and pedestrian routes and public transport networks, and access should discriminate in favour of environmentally friendly modes.
- Attention must be given to the most effective division of planning responsibilities between different levels of government planning and, where appropriate, wider powers granted to the (usually local) level of implementation.

Processes of European political and economic integration may work simultaneously for and against the promotion of such policies. Increased trade and competition for inward investment between regions and cities may serve to undermine arguments for integrated land use and transport policies. Not only will this result in higher CO_2 emissions, but also it will greatly detract from the social and economic benefits that such policies carry with them. Furthermore, despite a range of statements of commitment to environmentally sustainable development, the EU continues to promote economic growth without a sincere evaluation of the environmental impacts of some aspects of growth. The competition between parallel rationalities – global competitiveness and growth versus environmental sustainability – is represented in European policies, but it is the growth rationality that continues to dominate.

European level intervention, in the form of harmonization of planning regimes relating to travel inducing development and promotion of more energy efficient urban forms may prove essential to the effective implementation of planning related emission reduction strategies. European promotion of 'rights of access' by green modes (walking and cycling) to certain basic needs could also serve to catalyse the integration of land use and transport planning at local authority level.

While environmental protection and economic development are still, in practice, seen as conflicting interests, 'balanced' by planners with varying degrees of failure across Europe, the climate change issue promises that in the future, a convincing integration *must* be achieved. Local Agenda 21 provides both a framework and a degree of legitimate authority, at the most appropriate level of planning intervention, that may catalyse this integration. Successful integration will, however, require significant long term commitment. The concept of 'balance of interests', which has for so long amounted to a presumption in favour of economic development, is deeply entrenched in planning thinking.

The intensification of competition between regions may serve to reinforce this. Hence the implementation of policies, including land use planning policies, that seek to reduce CO_2 emissions related to the transport sector requires a high degree of political leadership and imagination. Politicians and other key decision makers will have to point to the ways in which economic opportunities can be framed (and indeed extended) by environmental priorities. The fact that most businesses and much of the population of Europe are currently reliant on (private) road transport should not disguise the fact that there is a widespread sense that the real interests of business and ordinary people are ill served by the current pattern of built form and high rates of mobility.

REFERENCES

Banister, D (1991) 'Energy use, transport and settlement patterns' Paper for Regional Science Association Annual Conference, Oxford, September

Blowers, A et al (1993) *Planning for a Sustainable Environment*, Earthscan, London

Breheny, M (1990) 'Strategic planning and urban sustainability', paper presented at the Town and Country Planning Conference on *Planning for Sustainable Development*, London 27–28th November 1990

City of Copenhagen, Lord Mayor's Department (1995) 'A tale of two cities: II', paper presented at conference on *The European City*, Copenhagen, April

Commission of the European Communities (1990) *Green Paper on the Urban Environment*, Commission of the European Communities, Brussels

Commission of the European Communities (1992) *The Future Development of the Common Transport Policy: A Global Approach to the Construction of a Community Framework for Sustainable Mobility*, COM (92) 494 final, Commission of the European Communities, Brussels

Countryside Commission (1995) *Climate Change, Acidification and Ozone*, Countryside Commission, Cheltenham

Department of the Environment (1988) *Major Retail Development, Planning Policy Guidance Note 6*, Department of the Environment, London

Department of the Environment (1994a) *Planning Policy Guidance Note 13 Transport*, HMSO, London

Department of the Environment (1994b) *Sustainable Development*, Cmd. 2426, HMSO, London

ECMT and OECD (1995) *Urban Travel and Sustainable Development*, OECD, Paris

Goodwin, P (1994) 'Sustainable development and transport policy', in Hope, C and Owens, O (eds) *Moving Forward: Obstacles to a Sustainable Transport Policy*, White Horse Press, Cambridge

Goodwin, P et al (1991) *Transport: The New Realism*, Report to the Rees Jeffries Road Fund, University of Oxford Transport Studies Centre, Oxford

Hardy, D (1991a) *From Garden Cities to New Towns: Campaigning for Town and Country Planning 1899–1946*, Spon, London

Hardy, D (1991b) *From New Towns to Green Politics: Campaigning for Town and Country Planning 1946–1990*, Spon, London

Hewett, J (1995) *European Environmental Almanac*, Earthscan, London

IPCC (1996) *Climate Change 1995: Impacts, Adaption and Mitigation of Climate Change*, Cambridge University Press, Cambridge

Kent County Council (1990) *The Transport Challenge: A New Approach for Kent*, Highways and Transportation Department, Kent County Council, Maidstone

Naess, P (1991) 'Environment protection by urban concentration', paper presented to conference on *Housing Policy as a Strategy for Change*, Norwegian Institute for Urban and Regional Research, Oslo

Netherlands Ministry of Housing, Spatial Planning and Environment, Ministry of Transport and Public Works and Ministry of Economic Affairs (1990a) *Working Document Guiding Mobility by a Location Policy for Businesses and Amenities*, Ministry of Housing, Spatial Planning and Environment, The Hague

Netherlands Ministry of Housing, Spatial Planning and Environment (1990b) *National Environmental Policy Plan Plus 1990–1994*, Ministry of Housing, Spatial Planning and Environment, The Hague

Netherlands Ministry of Housing, Spatial Planning and Environment (1994a) *National Environmental Policy Plan 2: The Environment: Today's Touchstone Summary*, Ministry of Housing, Spatial Planning and Environment, The Hague

Netherlands Ministry of Housing, Spatial Planning and Environment, Ministry of Transport and Public Works, Ministry of Economic Affairs (1994b) *Location Policy in Progress: The Story So Far*, Ref. 31220, Ministry of Housing, Spatial Planning and Environment, The Hague

Newman, P W G and Kenworthy, J R (1989) 'Gasoline consumption and cities: A comparison of US cities with a global survey' *Journal of the American Planning Association*, Vol 55, no 1, pp24–37

Owens, S (1986) *Energy, Planning and Urban Form*, Pion, London

Owens, S (1991) *Energy Conscious Planning*, Council for the Protection of Rural England, London

Owens, S (1995a) 'The compact city and transport energy consumption: a response to Michael Breheny', *Transactions of the Institute of British Geographers*, Vol 20, no 3, pp381–386

Owens, S (1995b) 'From "predict and provide" to "predict and prevent"?: pricing and planning in transport policy', *Transport Policy*, Vol 2, no 1, pp43–49

Owens, S and Cope, D (1992) *Land Use Planning Policy and Climate Change*, Department of the Environment, Planning Research Programme, HMSO, London

Peake, S (1996) Compilation and Synthesis of Policies and Measures to Address Climate Change in the Transport Sector, Unpublished report, IEA, Paris

Rees, J and Williams, S (1993) *Water for Life: Strategies for Sustainable Water Resource Management*, Council for the Protection of Rural England, London

RCEP (1994) *Transport and the Environment*, Eighteenth Report, Cmd. 2674, HMSO, London

SERPLAN (1989) *Regional Transport Statement*, RPC 1235, SERPLAN, London

Smith, J and Owens, S (1996) 'Planning against climate change: A review of land use planning strategies aimed at reducing transport-related greenhouse gas emissions', unpublished report, European University Institute, Florence

TEST (1989) *Trouble in Store: Retail Location Policies in Britain and West Germany*, TEST, London

Chapter 4 | # THE EU AND CLIMATE CHANGE POLICY: THE STRUGGLE OVER POLICY COMPETENCES

Ute Collier

Introduction

When the climate change issue was at the top of political agendas in 1990, EU environmental policy had just seen a successful decade with common action in a growing number of areas. Hence, there was optimism that the EU would also be able to implement a strong climate change strategy with common policy measures and to set an example to the rest of the world. Six years later, the EU's response to the climate change strategy lies in tatters.

Global agreements such as the FCCC have to be implemented through action at the national and local level, but action at the EU level can also be justified. In fact, the EU itself, as well as all the Member States separately, has signed the FCCC and thus is required to draw up an abatement strategy. Furthermore, the 15 EU countries constitute one of the most powerful economic blocks in the world (and account for nearly 15 per cent of global CO_2 emissions), so with a common strategy might be able to exert pressure on other countries to act on environmental issues. Additionally, it is useful to coordinate action between countries and to exchange information. Finally, certain measures for emission reductions, such as appliance standards or energy taxes, need harmonization to allow the functioning of the internal market.

The EU Commission made its first proposals for a climate change strategy in 1990 but, as later sections show, these have made limited progress. While there are various reasons for this policy failure (see Collier, 1996a), the lack of EU competence[1] in the energy area has been a major obstacle to the agreement of effective measures in the energy efficiency and renewable energy fields. Furthermore, one of the proposed instruments, the carbon/energy tax, suffered as fiscal measures have been notoriously difficult to agree at the EU level, with the Member States keen to guard their sovereignty in such matters. The debates about the appropriate level of action have intensified in recent years with an increased emphasis on the subsidiarity principle, and are likely to continue to influence EU climate change policy.

EMISSION CHARACTERISTICS OF THE EU

The generation of asymmetrical interests between Member States by environmental problems such as climate change has been identified as one of the obstacles to agreement on EU response strategies (Skjaerseth, 1994). As far as CO_2 emissions are concerned, there are indeed considerable variations in emission characteristics (both in terms of total and per capita emission characteristics) and abatement costs. As Figure 4.1 shows, Germany is by a large margin the largest emitter of total CO_2, followed by the UK and then Italy. Some of the smaller Member States, as to be expected, have very low emissions.

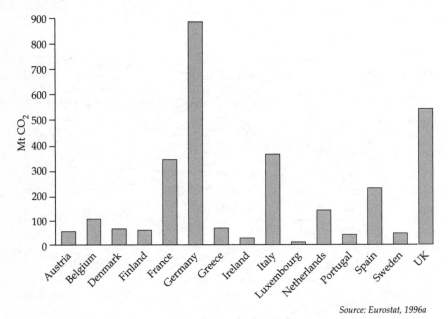

Source: Eurostat, 1996a

Figure 4.1: Total CO_2 emissions in the EU (1994)

---

1 The notion of competence is here used in terms of its legal meaning, ie allocation of responsibility.

Total emissions are obviously important, and reductions by the largest emitter countries are crucial in terms of the overall reductions that are seen as necessary to stem climate change. However, this means a very heavy burden on some countries and it is generally accepted that all industrialized countries will have to make a contribution to emission reductions. Some, such as France (see Chapter 8), have argued for taking into account per capita CO_2 emission levels when setting reduction targets, as these reflect differing national circumstances somewhat better. As Figure 4.2 shows, all Member States are well above the global average of just over one tonne per capita. One member state, Luxembourg, stands out way above the rest, due to large, inefficient steelworks, which, because of the country's small population, totally distort the per capita emission picture.

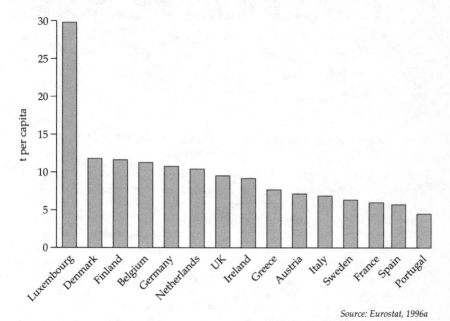

Source: Eurostat, 1996a

Figure 4.2: Per capita CO_2 emissions in the EU (1994)

Per capita CO_2 emissions vary for a number of reasons:

■ differing industrial structure (eg reliance on heavy industry);
■ stage of economic development;
■ proportion of non-fossil sources in energy requirements;
■ efficiency variations (of industrial processes, housing, car fleet, etc);
■ climatic differences (heating requirements);
■ urban structure;
■ modal split (public/private transport, road/rail transport).

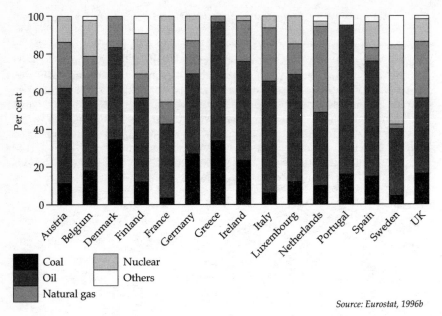

Source: Eurostat, 1996b

Figure 4.3: Gross inland energy consumption in the EU (1995)

Generally, the countries with high per capita emissions (eg Belgium, Denmark, Germany and the Netherlands) have a strong reliance on fossil fuels and a fairly energy intensive industrial structure. Also, because of relatively cold winters, they have high heating requirements. This applies even more to Sweden but it has heavily invested in building energy efficiency and hence has cut its heating requirements considerably. Furthermore, it has a high proportion of non-fossil fuel sources (mainly hydro and nuclear power). Meanwhile, Spain has low heating requirements and a lower level of economic development, but still the same level of CO_2 emissions as Sweden because of its heavy reliance on fossil fuels. Figure 4.3 shows these large variations in the contribution of different fuels to energy requirements in more detail.

As can be seen from Figure 4.3, Greece and Portugal are almost entirely dependent on two fossil fuels with a high carbon content: coal and oil. Denmark fares a little better with some input from natural gas. Natural gas plays a larger role in Italy, the Netherlands and the UK, but these countries are nevertheless highly fossil fuel dependent. France, Finland and Sweden are the only three countries with a share of non-fossil fuels of more than 30 per cent. As far as the non-fossil fuel share is concerned, the largest proportion is due to nuclear energy, although Austria, Sweden and Finland also derive a considerable amount of their energy requirements from renewable energies, in particular hydropower and biomass.

These differences in fuel characteristics have an important bearing on climate change policy and the prospects of finding a common EU approach to the issue. Countries with a greater reliance on coal and oil generally have more scope for emission reductions, as fuel switching in the electricity sector is one of the 'easier' ways to achieve them, especially switching to gas fired power stations which, as shown in Chapter 2, are currently economically very attractive. These are options that countries such as France and Sweden do not have (see Chapters 8 and 10). Furthermore, energy efficiency measures, in particular those in the electricity sector, will have less effect on emissions the lower the carbon intensity of the energy system is. If there are flat emission targets, these countries will have to put much more emphasis on the transport sector, which is generally considered more difficult.

Furthermore, other diverging factors such as different levels of economic development can make finding a common approach difficult. The poorest member states in the EU, known as the cohesion countries (Greece, Ireland, Spain and Portugal), have always argued for special dispensation on this basis. The likely impacts of climate change also vary. Rotmans et al (1994) demonstrate that the Southern countries can expect adverse effects, mainly due to sea level rises and water shortages. Conversely, the Northern Member States may benefit as a result of increased agricultural yields. As a result, the costs and benefits of any policy action will vary.

Overall, because of the factors discussed above, a common EU approach to the climate change issue faces a number of potential obstacles. As such this divergence in national circumstances is nothing new and has had to be dealt with in many other EU environmental decisions, including the large combustion plant directive, where differentiated emission reduction targets for SO_2 and NO_x were set for different countries. The precedence to reach an agreement is thus available. However, two additional, and closely related, obstacles have made themselves felt in the climate change case; the lack of EU competences in the energy area and the growing emphasis on the subsidiarity principle.

SUBSIDIARITY AND THE LACK OF EU COMPETENCES IN ENERGY POLICY

As Golub (1996) has argued, the concept of subsidiarity has always existed in EU environmental policy, although until recently this has been mostly implicit; with the first Environmental Action Programme (EAP), for example, referred to five possible levels of action and stressed the need 'to establish the level best suited to the type of pollution and to the geographical zone to be protected'. The Single European Act (SEA) for the first time incorporated provisions which would later become known as the subsidiarity principle, but only in Article 130r covering environmental policy:

> *The Community shall take action relating to the environment to the extent to which the objectives … can be attained better at Community level than at the level of individual Member States (SEA, 1987).*

In practice, these provisions had little influence on the development of EU environmental policy prior to the Treaty on European Union (TEU); environmental proposals actually reached record numbers and several important directives were adopted post SEA (Collier, 1996a). However, during the discussions surrounding the Maastricht Treaty, the subsidiarity issue was suddenly pushed to the top of the political agenda, not only in relation to environmental policy but also for EU policy making as a whole. The TEU then included the following requirement:

> *In areas which do not fall within its exclusive competence, the Community shall take action ... only if and in so far as the objectives of the proposed action cannot be sufficiently achieved by the Member States and can therefore, by reason of the scale or effects of the proposed action, be better achieved by the Community (Council of the European Communities, European Commission, 1992).*

As signature of the Climate Change Convention obliges the Member States to draw up emission reduction programmes, within the framework of subsidiarity this could be considered sufficient to ensure the achievement of the policy objectives, without a need for EU level measures. However, there are several arguments to justify EU level action. Firstly, some instruments such as taxes or standards have to be EU wide to avoid distortion. Secondly, an EU target allows some flexibility to differentiate reduction targets between Member States and aim at a cost-effective approach to meeting the overall target. Furthermore, if the EU, which accounts for 15 per cent of global emissions, acts as a block, it might be able to exert pressure on other countries to act. Additionally, as Krämer (1995) has argued, action at EU level is often the only way to ensure that environmental measures are taken in all Member States, considering the lack of concern in some countries. While it is easy to find a justification for at least some EU level action on climate change, the whole subsidiarity debate has been politicized, so that in the end decisions are made on the basis of political and economic priorities, rather than on environmental ones (Collier, 1996b). However, the subsidiarity debate might acquire another dimension in the future as the very fact that the EU is a party to the FCCC raises the issue of competence and the allocation of responsibilities between the EU and its Member States. As Macrory and Hession (1996) argue, this has not been a problem to date given the open ended nature of the Convention, but such legal issues may well become important if and when a binding protocol is agreed. In this case, the substantive requirements of EU law would increase and the debate about the division of competences would be likely to intensify, especially if qualified majority voting (QMV) became more extensive.

A specific problem in the subsidiarity context is that the EU's competence in the crucial energy field has been limited and, in a climate of subsidiarity, it is unlikely that its role will be extended in any substantial way. While the first of the European Communities, the European Coal and Steel Community, had an energy source at its heart, the Treaty of Rome made no separate mention of an energy policy. Efforts to draw up a common energy policy date back to 1962 when a working party on energy adopted a memorandum on energy policy, designed to achieve the free circulation of energy within the common market (McGowan, 1991). In subsequent years, especially during the oil price shocks, there were a number of attempts to define common energy policy guidelines, but as these were entirely voluntary they had little influence on developments. During the discussions on the TEU, there were efforts by the Commission to include a chapter on energy policy but this failed (Collier, 1994). Article 3 of the TEU thus allocates to the EU responsibility for *measures* in the energy area rather than a *policy*; an important distinction. Furthermore, under article 130s qualified majority voting is to be used for environmental policy, but this article explicitly excludes measures 'significantly affecting a Member State's choice between different energy sources and the general structure of its energy supply'.

As the 1996 Intergovernmental Conference (IGC) approached, the inclusion of an energy article in the TEU was again called for by the Commission, but it seems unlikely that this call will be accommodated. This does not mean that the EU has absolutely no powers in the energy area. Apart from provisions under the European Coal and Steel Community and Euratom, a whole range of common measures exist. A recent listing in a Commission publication[2] contains a total of 109 measures (regulations, directives and decisions). However, these do not amount to a comprehensive policy and, effectively, Member States have managed to retain their sovereignty in all crucial areas of energy policy. As later sections show, efforts have been made to draw up policy measures in the areas of energy efficiency and renewables in response to the climate change issue, but their success has been somewhat limited and clearly influenced by the subsidiarity debate.

Meanwhile, one energy area to which the Commission and some Member States have given priority in recent years, liberalization, is not necessarily beneficial in environmental terms. Here efforts have focused on the achievement of the Internal Energy Market (IEM) through a liberalization of the energy sector, for which the Commission presented its proposals in 1992 (European Commission, 1992a). A main focus of the proposals is the introduction of third party access[3] (TPA) to the closed national energy markets, with the ultimate aim of lower energy prices, and hence greater competitiveness, for large industrial consumers (Collier, 1994). The IEM proposals encountered much opposition, especially from France which has been very keen to protect the position of its public monopoly, EdF. New proposals were made in 1993 which authorized Member States to impose

---

2 *Energy in Europe*, December 1994, pp 36–49.
3 In principle, TPA is an important issue environmentally, as it means independent producers can sell electricity to the grid or the consumer directly. The opening of the market to independent producers can be to the benefit of renewable energies and CHP, which in the past have been hindered by unfavourable (or no) market access conditions. However, most of the EU discussion has been much narrower, allowing large industrial consumers to buy their electricity wherever it is cheapest, which may or may not involve independent producers.

public service obligations (which may include environmental protection) on the electricity sector (European Commission, 1993b). Furthermore, the French suggested the idea of a single buyer model[4] instead of TPA. A further working paper issued in 1995 suggests special access arrangements for renewables and CHP plants. In June 1996, finally, a common position was reached in the Council which implied that the single buyer and TPA models will exist in tandem. Initially, access will only be for large consumers with a yearly consumption exceeding 100 GWh, which effectively means that just 22.6 per cent of the market will be opened up. This figure will increase to one third by 2003.

The whole IEM discussion has proceeded without much reference to the climate change issue. While, as Chapter 2 shows, there are clearly beneficial changes that could be made to the EU energy markets, these are unlikely to be achieved by liberalization per se. The whole process is indicative of a continuing lack of integration between energy and environmental policy decision making at the EU level (Collier, 1994; European Commission, 1995a). DG XVII finally commissioned a report on the environmental implications of the IEM (focusing on its effects on CHP) in 1995, the results of which are unlikely to play any significant role in further developments. Meanwhile, as the next sections show, progress has been slow as far as specific policy measures on climate change are concerned.

ATTEMPTS AT DEFINING A COMMON CLIMATE CHANGE STRATEGY

Early Policy Developments

The climate change issue hit the policy agenda in the EU at what could be seen as the 'heyday' of EU environmental policy and, possibly, European integration in general. Some important directives on air pollution were agreed in the late 1980s, on emissions from large combustion plants and on emissions from cars. More importantly, in 1987 EU action in the environment area was finally given a separate legal basis by inclusion in the SEA. The TEU in principle further strengthened environmental policy by introducing QMV to environmental policy and by making 'sustainable growth respecting the environment' a general objective of the EU. The latter in fact reflects a general move towards a more comprehensive approach to dealing with environmental problems.

The EU has made 'sustainability' the main theme of its Fifth EAP. It also specifically stresses the importance of the concepts of subsidiarity and shared responsibility in achieving 'sustainability', namely through the mixing of actors and instruments at the appropriate levels. The Commission and the Member States increasingly have recognized the importance of the

---

4 This implies that independent producers are not allowed to sell electricity directly to the consumer. Instead, the electricity utility acts as a 'single buyer' and then resells the electricity to the consumer.

principle of policy integration, which was first mentioned in the SEA and then reiterated in the TEU. Indeed, in the early 1990s the Commission produced discussion papers on energy and transport policy, the most relevant areas to climate change, presenting its thinking on how this integration could be achieved. However, as mentioned above, this has not really happened in the energy field. Additionally, in fiscal policy efforts at the harmonization of various taxes have essentially failed.

The development of EU climate change policy has already been described in some detail elsewhere (see eg Jachtenfuchs, 1994; Haigh, 1996; Skjaerseth, 1994; Wynne, 1993), so only the most important developments are outlined here. The EU response to the issue began with a Commission Communication to the Council in 1988 (European Commission, 1988). This was essentially a stocktaking exercise, summarizng the greenhouse science and the outcome of meetings such as the Toronto Conference. An ad hoc committee was established in the Commission in 1989, including a total of ten Directorates-General (DGs) likely to be most affected by the development of an EU climate policy. Subsequently, DG XI (Environment), DG XVII (Energy) and DG XXI (Indirect Taxation) became the most important players in the policy process. In October 1990, the Energy and Environment Council of Ministers met for the first time in a joint session just before the Second World Climate Conference. The main purpose of the joint council was to agree a CO_2 reduction target for the Community so that a unified stance could be presented at the Conference. At the time, some Member States had already set CO_2 targets, although at different levels, while others had no targets at all.

The aim was to return emissions to 1990 levels by the year 2000 for the EU as a whole, while accommodating the less industrialized countries' growth requirements. This accommodation applied to Greece, Ireland, Portugal and Spain, but the UK also refused to move from its target date, which at that time was 2005. Some Member States (namely Denmark, Germany and the Netherlands) had already committed themselves to stricter targets, which allowed scope for other Member States to meet less stringent targets increasing their emissions or stabilizing later. Wynne (1993) thus calls the stabilization target an 'ambiguous supranational concoction'. Initially, there was talk of equitable target sharing, that is allocating individual targets for CO_2 emissions to the Member States, according to their development needs. However, as Grubb and Hope (1992) point out, attempts to reach agreement on sharing the target never really got off the ground.

Discussions on the CO_2 target were relatively easy compared to the subsequent discussions about drawing up a CO_2 strategy. Disagreements on a number of issues emerged, with different viewpoints from the various DGs involved. In the Council, there was animosity to the proposals from some Member States. Spain, for example, felt it was too early for an EU programme on climate change, while Portugal did not consider the issue a problem at all (Collier, 1994).

Originally, a range of specific measures was envisaged for the strategy which, apart from a carbon tax, included a variety of efficiency standards (buildings, water heaters and cookers), a speed limit of 120 km/h, implementation of least cost planning principles and measures to promote waste recycling.[5] However, consecutive drafts saw a significant scaling down of the proposals. The Commission eventually published a first Communication on the issue in 1991 (European Commission, 1991) and a second in 1992 (European Commission, 1992b).

The Community strategy to reduce CO_2 emissions

The 1992 Communication, entitled 'A Community strategy to limit carbon dioxide emissions and to improve energy efficiency', firstly outlined the Commission's proposals for the strategy in general terms. It was accompanied by proposals for four specific measures as follows:

■ a framework directive on energy efficiency [Specific Actions for Vigorous Energy Efficiency (SAVE)];
■ a directive on a combined carbon/energy tax;
■ a decision concerning the specific actions for greater penetration of renewable energy resources (ALTENER);
■ a decision concerning a mechanism for the monitoring of Community CO_2 emissions and those of other greenhouse gases.

Furthermore, the Commission's energy technology support programme (THERMIE) was expected to contribute to emission reductions. While the 1991 Communication had assumed the need to reduce emissions by 11 per cent from the 1990 level to achieve stabilization (European Commission, 1991), the 1992 Communication revised this figure upwards to 12 per cent due to an accelerating growth in emissions in 1991 (European Commission, 1991). The different measures and programmes were expected to contribute different proportions of the required reductions as shown in Table 4.1.

Table 4.1. *Projected Emission Reductions from the EU Climate Change Strategy*

Proposed measures for stabilization	Expected CO_2 reduction
Carbon/energy tax (and accompanying national measures)	6.5%
SAVE	3.0%
THERMIE	1.5%
ALTENER	1.0%
Total	12.0%

Source: European Commission, 1992b

The carbon/energy tax and the SAVE programme were thus expected to achieve the bulk of the emission reductions. Furthermore, work started in

---

5 According to a draft communication of 28th November 1990, detailing the initial proposals.

the Commission on measures to reduce the emissions from the transport sector and proposals were developed to improve demand side planning in the energy sector. However, by 1996 little progress had been made with these measures or the main elements of the strategy:

- the proposal for a combined carbon/energy tax had been blocked;
- the SAVE programme on energy efficiency had been turned into a framework directive, with doubts about its effectiveness;
- the ALTENER programme on renewables continued to be underresourced and mainly consisted of non-binding targets;
- proposals for reducing CO_2 emissions from cars were delayed due to disagreements within the Commission[6] and appeared unlikely to make progress;
- a proposal for rational resource planning in the energy sector was delayed until 1995 and continues to face much opposition from the energy sector.

Considering the promising start of the climate change discussions in the late 1980s, questions arise as to what went wrong. The following sections examine some of the components of the strategy in more detail, identifying the problems encountered.

Policy Failure: the Carbon/Energy Tax

A main focus in the development of the Commission's strategy on CO_2 was on the possibility of introducing a tax in order to internalize some of the external costs of energy. Pressure for an EU level tax came from the fact that three Member States (Denmark, Germany and the Netherlands) were threatening to introduce carbon taxes unilaterally, thus infringing the Commission's attempt to harmonize taxes for the proper functioning of the single market. The tax also fitted in with a general growth in interest in market based instruments to achieve environmental objectives (see Collier, 1996a). However, considering the EU's previous[7] difficulties in the fiscal policy area, it was clear that this was not going to be a measure on which agreement would be reached easily.

First proposals for a tax were put forward in a Communication to the Council in late September 1991 (European Commission, 1991). It was decided that there should not be a single CO_2 levy as this would favour nuclear power, which a number of Member States opposed. The proposed tax thus was a so-called hybrid carbon/energy tax amounting to $10/barrel by the year 2000, starting with $3/barrel as of 1 January 1993 and increasing by $1 annually. Further details were put forward in a communication in

---

6 Jacques Delors' Commission finally published the proposals in late 1994 [*COM* (94) 647] and left follow up to its successors. It is not yet clear as to whether the new Commission will pursue any of the options mentioned.
7   During the 1980s there were long discussions about the harmonization of value added tax (VAT) which resulted in a compromise agreement which meant little change for most Member States.

June 1992 (European Commission, 1992c). The first casualty was the proposed starting date which by the time of publication was no longer referred to.

According to the 1992 proposals, half of the tax was to be based on CO_2 emissions (expressed in tonnes) and half on the calorific value of the fuel (expressed in gigajoules). Energy from renewables (except hydropower plants above 10 MW) was to be exempted from the tax. From the start, it was clear that the tax proposals would attract opposition from various industrial groups. Intensive lobbying against the tax by these duly took place, accompanied by threats to move industrial production outside the EU. As a result, a number of concessions were made which would have substantially weakened the effect of the tax.

Firstly, Member States would have been authorized to grant tax reductions of up to 75 per cent to firms whose energy costs amount to at least 8 per cent of the value added of its products and whose competitiveness might be threatened by the tax (European Commission, 1992c). Member States would also have been allowed to grant temporary total exemptions to firms that had embarked on 'substantial efforts to save energy or to reduce CO_2 emissions'.[8] This vague statement was liable to lax interpretation and the exemptions seriously compromised the effectiveness of the tax, as they meant that the largest consumers of energy in the EU would have paid the lowest rates of tax, thus giving them little additional incentive to invest in energy efficiency.

Despite these concessions, the proposals made little progress when discussed at various environment and ECOFIN (Economic and Finance) council meetings. Because this was designated as a fiscal measure, unanimous agreement had to be achieved, but this was not forthcoming. The main objection came from the UK,[9] which was vehemently opposed to any European intervention in tax matters. As no progress was evident, various new approaches were discussed. These included a possible reform and harmonization of current energy taxes. However, no agreement could be reached on this either. At the 1994 Essen summit, Spain, Portugal and Luxembourg joined the UK in a declaration stating that 'the assessment of the need to introduce a tax must remain within the competence of each Member State'.[10]

The Commission was asked to submit a new proposal to outline common guidelines for those Member States who wanted to implement their own taxes. New proposals were issued in 1995, the wording of which effectively meant that the tax would be voluntary during a transitional period until 2000 but binding thereafter.[11] This has resulted in renewed opposition by industrial lobby groups and has left the Member States divided.[12] Currently, five Member States (Austria, Denmark, Finland,

8 Additionally, the Council, acting unanimously on a proposal from the Commission, could have suspended the application of the tax in 'exceptional cases in order to take account of the special situations in Member States'.
9 Although, according to one official in the Commission, some other Member States were also opposed but quite content to let the UK assume the role of the 'bad guy'.
10 'Council defines strategy to reduce CO_2 emissions', *Agence Europe*, 17th December 1994, p 9.
11 *Europe Environment*, No 455, 23rd May 1995.
12 *COM* (95) 172.

Netherlands and Sweden) have introduced CO_2 taxes, but it is unlikely that any other Member States will follow suit. In most cases, the taxes applied focus on the domestic sector, while the industrial sector receives generous or even total exemptions. In any case, it is not clear how effective such a tax would be in a climate that forces down energy prices.

Subsidiarity has been used as one of the justifications for opposing the carbon/energy tax. The UK has been at the forefront of this opposition and has argued that it would be more appropriate to develop a tax at the national level (Collier, 1996c). After its decision to impose VAT on domestic fuel in March 1993, the UK government claimed that it had already instituted a form of carbon tax. However, as Golub (1994) has shown, Britain's concern about sovereignty and national interests previously had been influential in EU environment policy and there is little doubt that a main reason for objecting to the tax was not a true belief in subsidiarity, but a general reluctance to surrender decision making powers to the EU, especially on important matters such as taxes. At the same time, other Member States, in particular the cohesion countries, were concerned about the effect of the tax on their competitive position and favoured subsidiarity arguments. Furthermore, France was not satisfied and wanted a pure carbon tax, in order to protect its nuclear industry. Effectively, these Member States have been able to 'hide' their opposition behind the UK. Generally, it can be said that there was very little support for the proposals (except from some smaller Member States, in particular Denmark and the Netherlands) which, without any strong advocates, could not progress. Other proposals under the 1990 climate change strategy have not fared much better.

Limited Success: The SAVE Programme

Energy efficiency has long been a concern in EU energy policy discussions, although action has essentially restricted itself to the setting of non-binding targets, with some small supporting R&D programmes. Initially, SAVE was conceived to enable the EU to meet its 1995 energy objective of improving efficiency by at least 20 per cent (European Commission, 1986), the achievement of which was beginning to look increasingly unlikely in the early 1990s. First proposals for SAVE were published under separate cover in November 1990 (European Commission, 1990) and were subsequently revised and integrated into the CO_2 strategy (European Commission, 1992d). The proposals of November 1990 envisaged a variety of measures under SAVE but, as ever, they were watered down substantially. Most significantly SAVE has been turned into a so-called framework directive, which means that the EU only sets the general principles for action, on which Member States then have to base their programmes of measures. This has been a direct result of Member States invoking the principle of subsidiarity.

The final directive thus states that the Member States shall draw up and implement programmes in six areas, as follows:

■   minimum insulation standards for new buildings;

■ energy certification of buildings;
■ billing of heating costs based on actual consumption;
■ promotion of third party financing for public sector investments;
■ inspection of boilers;
■ energy audits for businesses with high energy consumption.

However, the Member States essentially have a free hand in designing and implementing such programmes. The directive states that:

> *Programmes can include laws, regulations, economic and administrative instruments, information, education and voluntary agreements whose impact can be objectively assessed (European Commission, 1993a).*

The targets and timescales suggested in the 1992 proposal were abandoned. These would, for example, have required the certification of public sector buildings at a rate of at least 5 per cent of the existing stock per year (European Commission, 1992d), while the final directive merely talks generally of the need for an energy certification of buildings. Furthermore, the inspection of heating installations is now restricted to those above 15 kW and energy audits are only required for industrial undertakings with high energy consumption rather than for businesses in general. The Member States have to report to the Commission every two years on the results of the measures taken, so the effectiveness of SAVE will only emerge in a few years' time. The Commission itself has stated that the high degree of flexibility left to Member States renders the estimation of the effects of SAVE highly uncertain (European Commission, 1994). The UK, for example, has made it known that it sees no need for any further legislative measures as a result of the SAVE directive,[13] yet (as Chapter 6 discusses in more detail) the UK's activities in the energy efficiency area are totally inadequate.

A proposal for SAVE II was presented by the Commission which aimed at energy savings of 60–70 Mtoe per year by the year 2000 and involved a budget of 150 million ECU between 1996 and 2000. However, at the Energy Council in May 1996, France, Germany and the UK refused to approve a budget any higher than 45 million ECU[14] which makes SAVE II unlikely to be any more effective than SAVE I.

Another measure currently under discussion which potentially offers emission savings is for a directive on energy efficiency requirements for electric refrigerators and freezers, which was initially proposed by the Commission in December 1994. The minimum standards to be set by the directive were initially weakened by the Council, then scaled up by the European Parliament, with the Directive finally adopted by the Council on a compromise level in July 1996, requiring initial energy efficiency improvements of 15 per cent (compared to average consumption rates) within 3

---

13 As stated in an explanatory memorandum dated October 1993 on the SAVE directive, submitted by the Department of the Environment to the House of Commons.
14 *Agence Europe*, 6th May 1996, p 3.

years (European Commission, 1996a). While this level is far below that which the best appliances in the market already achieve, appliance manufacturers lobbied Brussels heavily to renounce the proposals in return for voluntary agreements. While this was not successful (despite the directive), in general EU level action on energy efficiency is set to remain weak. This also applies to the transport sector, where emissions are particularly fast growing and yet the car industry is lobbying heavily against the imposition of any fuel consumption standards.

Finally, one small but notable success of the influence of the climate change discussions on EU energy policy decision making has been the fact that a framework energy labelling directive, which was first proposed in 1979, was finally agreed in 1992. Subsequently, an application directive for domestic refrigerators was adopted in 1994 and one for ovens in 1995. While labelling is obviously important in dealing with the information deficit, one of the obstacles to greater energy efficiency investments, these directives will only result in minor efficiency gains, while none of the other obstacles are being dealt with.

ALTENER: A Very Modest Renewables Programme

As Table 4.1 indicates, the ALTENER programme for renewable energies was going to be the weakest instrument in the climate change policy. However, it now seems uncertain that even the projected 1 per cent reduction in CO_2 emissions can be achieved. Renewable energies have featured in EU energy discussions since the mid 1980s, but it was not until 1993 that specific numerical targets were adopted in conjunction with the ALTENER programme (Grubb, 1996). According to this, in order to reduce CO_2 emissions by 180 Mt by 2005, the following need to be achieved (European Commission, 1993b):

■ Increasing renewable energy sources' contribution in the coverage of total energy demand from 4 per cent in 1991 to 8 per cent in 2005.
■ Trebling the production of electricity from renewable energy sources (excluding large hydropower facilities).
■ Securing for biofuels a market share of 5 per cent of total fuel consumption by motor vehicles.

To assist the achievement of these objectives, 40 million ECU of EU funding has been allocated for the first five years (1995–2000), mainly to be used for various pilot studies. Considering Grubb (1996) has estimated that to achieve the electricity sector target alone would require the redirection of over 20 billion ECU of investment from 1995 to 2005, this sum appears very modest, in particular in view of the current low energy prices and the failure to agree on the carbon/energy tax. On the positive side, both the reform of the Common Agricultural Policy and the increased funding for the Structural Funds mean that there are now important extra funding possibilities for investments in renewable energies, in particular in the cohesion countries (see for example, Chapter 9 on Spain).

Rational Resource Planning

An additional measure which would result in CO_2 reductions, if adopted, concerns rational resource planning in the gas and electricity industries, an idea which was first discussed in the Commission in 1991. However, the proposal was not published until September 1995, mainly because of opposition by the Industry Commissioner, Martin Bangemann, who was concerned about the compatibility of the proposal with parallel proposals for energy market liberalization.

The proposal draws on the US experience with Integrated Resource Planning (IRP), which obliges energy companies to consider Demand Side Management (DSM) when planning new capacity needs. The directive would require Member States to establish procedures whereby electricity and gas distribution companies have to periodically present integrate resource plans to the competent authorities. Furthermore, Member States would be expected to review existing legislation to ensure mechanisms are established which permit the companies to recover expenditure on energy efficiency programmes (European Commission, 1995b).

However, it is not clear how rational planning could operate in an increasingly competitive market, as the US experience applies IRP to companies which operate as regional monopolies. The introduction of competition in California, for example, is now threatening IRP and DSM programmes (Sioshini, 1996). The Commission itself has been rather vague as concerns the compatibility between this directive and the IEM. Furthermore, opposition from the energy lobby, in particular the electricity association Eurelectric, continues and it is somewhat doubtful whether this proposal will pass through the Council.

RELIANCE ON NATIONAL AND LOCAL ACTION

As the effect of the EU level measures is not going to be adequate enough to achieve the initially set targets for emission reductions, the onus for action invariably has to be at the national and subnational levels. National governments have clearly an important role in climate change abatement in a number of areas including:

■ setting overall energy and transport policy frameworks;
■ regulatory frameworks for the energy sector (including energy price control);
■ energy taxation;
■ energy efficiency policies (especially standard setting);
■ renewable energy policies;
■ support for energy R&D.

The following chapters analyse in some detail how policies are developing at the national level in the six case study countries. As far as all 15 Member States are concerned, both the targets set initially and the progress with achieving them differ considerably, as Table 4.2 shows.

Table 4.2. *Member State Emission Targets and Projected Emissions for 2000*

State	Target	Member State projections	Modified Commission trajectory
Austria	20% reduction by 2005 (1988 base)	0.6	8
Belgium	5% reduction by 2000 (1990 base)	−1.1	3
Denmark	20% reduction by 2005 (1988 base)	−11.9	7
Finland	Stop emission growth by 2000	29.7	33
France	Stabilization at less than 2 tC per capita	8.5	13
Germany	25% reduction by 2005 (1990 base)	−13	−10
Greece	'Realistic objective' +15% between 1990 and 2000	14.3	19
Ireland	Increase by 2000 limited to 20% (1990 base)	20.5	25
Italy	Stabilization by 2000 (1990 base)	2.9	6
Luxembourg	Stabilization by 2000 (1990 base)	−24.1	−20
Netherlands	3% reduction by 2000 (1990 base)	−0.4	10
Portugal	Increase by 2000 limited to 40% (1990 base)	36	36
Spain	Increase by 2000 limited to 25% (1990 base)	20.8	23
Sweden	Stabilization by 2000 (1990 base)	4.4	6
UK	Stabilization by 2000 (1990 base)	−6.1	−2
EU15	Stabilization by 2000 (1990 base)	−1	3
	(DG XVII projection:		5.4)

Source: European Commission, 1996b

In principle, the addition of national projections shows the level of total expected CO_2 emissions in 2000. One has to be very careful with the projected figures as they include a high level of uncertainty, not only concerning issues such as economic growth or fuel prices, but also in terms of the effects expected from measures in national climate change strategies. According to the European Commission (1996b), there is generally insufficient information about these, with the expected impact of individual measures or sets of measures not being quantified and not enough information being available as to the state of implementation of many of them. Owing to various uncertainties in the projections by the Member States, the Commission has made a modified projection using common assumptions, for example regarding growth rates and fossil fuel prices, to arrive at more comparable figures. These figures, shown in the final column, arrive at an emission increase of 3 per cent.

Notwithstanding the data problems, Table 4.2, supported by other data (eg IEA, 1995; Climate Action Network, 1995) indicates that most countries are unlikely to reach their targets. Few countries actually project emission reductions and, where they do, these are incidental and likely to be short lived (see Chapters 5 and 6 on Germany and the UK), unless effective policy measures are implemented. Luxembourg appears to be making the greatest progress, due to a restructuring in its steel industry, again not for climate change reasons (European Commission, 1996b). The data for recent years have been mildly encouraging, and in 1994 CO_2 emissions were 2.7 per cent lower than those in 1990.

Should the EU reach its stabilization target in 2000, national climate change programmes will clearly only be able to take part of the credit. Another part, even if possibly quite small, will be due to the efforts of local authorities in this field. While the subsidiarity debate has been couched in terms of EU versus national level action, the relevance of lower levels of action must not be forgotten. Local authorities in the EU have considerable influence over a number of activities which are important for CO_2 emissions, especially in the energy and transport fields. The scope for action depends to some extent on the exact nature of local authority competences, which vary from country to country. As far as energy is concerned, the potential influence is obviously greatest where there are municipally owned energy companies (eg in Germany, Austria and Sweden) which can directly influence investments and pricing and thus impact on emissions, provided the right priorities are being set. Additionally, in all countries local authorities are important as energy consumers themselves, for example in terms of streetlighting and public buildings. Furthermore, the issue of energy consumption clearly has wider connotations and is intricately linked with general societal consumption behaviour and mobility requirements. Efforts to reduce energy consumption thus have to be linked to other issues such as re-use, recycling and public transport provision, which are generally planned for and provided at the local level. Finally, land use planning decisions for the siting of housing, shopping and industrial areas are taken at the local level and have a crucial impact on mobility needs and hence energy consumption.

Both the EU and national governments have largely ignored the local level as relevant when drawing up climate change strategies. Yet, at least in some Member States, local governments have been very dynamic, often with greater initiative than their national government. At the European level, currently two organizations are attempting to encourage local authorities to reduce their CO_2 emissions; the International Council for Local Environmental Initiatives (ICLEI) and the Germany based Climate Alliance (Klimabündnis). ICLEI runs a 'Cities for Climate Protection' campaign which encourages towns to sign up to reduce their CO_2 emissions by 20 per cent by 2005, compared to 1987 levels. The aim is to enlist the participation of at least 100 cities worldwide (by November 1996, there were over 150 members) whose combined CO_2 output represents 5 to 10 per cent of global emissions. The Climate Alliance aims at a reduction by 50 per cent of per capita CO_2 emissions by the year 2010, compared to 1987 levels. Around 400 municipalities (many of them relatively small) from 10 countries have signed the pledge, nearly half of them based in Germany (Klimabündnis, 1995). Similarly to the ICLEI pledge, municipalities are expected to draw up emission inventories and programmes for emission reductions. One problem in achieving any type of concerted action at the local level is that the scope for action varies tremendously between EU Member States. While in some countries local authorities have extensive powers and resources in these areas, in others they have relatively little scope for action.

To date, there is little information about these local activities, which makes it difficult to assess their likely impact on CO_2 emissions. There is some dissemination of information through ICLEI and the Climate Alliance, and there is some EU level support for local activities. DG XI (Environment) financially supports a 'sustainable cities' programme, DG XVII (Energy) supports 16 projects under its 'Regional and Urban Energy Programming' initiative, while DG XII (Research) has funded a research project on the optimization of climate change strategies in EU municipalities. However, these are all very small scale projects and, overall, there remains a general lack of coordination between the different levels of action in the climate change area. The Climate Alliance feels that for effective local action a greater delegation of powers to the local level is needed everywhere (Klimabündnis, 1993). As there will never be a uniform allocation of powers to the regional or local levels in the EU, it is very difficult to achieve an effective policy exclusively through subsidiary action at these levels.

CONCLUSIONS: WHAT FUTURE ROLE FOR THE EU?

As this chapter shows, the attempt to implement a common strategy to deal with the climate change issue essentially has been a failure, with only some relatively insubstantial EU level measures in existence. The onus for action is thus on the Member States, with policies and measures needed at both regional and local levels. In principle, there is no compelling reason to implement inflexible common policy measures for climate change. There is no real need for every member state to have the same type of energy efficiency policies or prescribe the use of certain technologies. In fact, if all Member States took their obligation under the FCCC seriously, EU action could almost be restricted to coordination and funding for research, development and demonstration. However, in reality, environmental issues appear to receive little attention in a number of Member States. Without measures being enforced by the EU, they might well take no action at all.

Furthermore, in the case of a CO_2 tax some useful lessons could be learnt from the application of different tax levels, as there is uncertainty as to which tax level is most appropriate for achieving emission reductions. However, concerns about economic competitiveness and the functioning of the internal market mean that such measures have to be agreed at EU level with common denominators. In legal terms, the participation of the EU in the FCCC, in particular in view of a future protocol, means that some degree of EU level measures may become necessary to ensure compliance.

Considering the continuing need for a least some types of CO_2 reduction measures to be implemented at EU level, questions arise as to the scope for further policy developments. In March 1995, in preparation for the first Conference of the Parties of the FCCC, the Commission presented, on request by the Council, a working paper on the current state of policy developments, as well as options for the period 2005–2010. In general, the paper states that a considerable political commitment is required in various policy areas (principally energy, transport, fiscal etc) if CO_2 limitations are to take place effectively. The suggestions for action taken include some very broad options such as changing market structures, removing barriers to energy efficiency and renewables and the integration of environmental concerns in the fiscal system (European Commission, 1995c). Currently, the Commission is reconsidering ideas about 'burden sharing' or, to use the preferred phrase, differentiated target setting, with or without further common measures. Considering the history of opposition to measures to date, EU climate change policy is unlikely to progress smoothly.

REFERENCES

Climate Action Network (1995) *National Plans for Climate Change Mitigation: Independent Evaluations*, Climate Action Network, Brussels

Collier, U (1994) *Energy and Environment in the European Union: the Challenge of Integration*, Avebury, Aldershot

Collier, U (1996a) 'The European Union's climate change policy: limiting emissions or limiting powers?', *Journal of European Public Policy*, Vol 3, no 1, pp123–139

Collier, U (1996b) *Deregulation, Subsidiarity and Sustainability: New Challenges for European Union Environmental Policy*, Working Paper 96/60, Robert Schuman Centre, European University Institute, Florence

Collier, U (1996c) 'Developing responses to the climate change issue: the role of subsidiarity and shared responsibility' in Collier, U, Golub, J, and Kreher, A (eds) *Subsidiarity and Shared Responsibility: New Challenges for EU Environmental Policy*, Nomos Verlag, Baden-Baden

Council of the European Commission (1992) *Treaty on European Union*, Office for Official Publication of the European Communities, Luxembourg

European Commission (1986) 'Council resolution concerning new Community energy policy for 1995 and convergence of the policies of the Member States', *Official Journal of the EC*, C 241, pp1–3.

European Commission (1988) 'The greenhouse effect and the Community', *COM* (88) 656 final

European Commission (1990) 'Proposal for a Council Directive concerning the promotion of energy efficiency in the Community', *COM* (90) 365 final

European Commission (1991) 'A Community strategy to limit carbon dioxide emissions and to improve energy efficiency, Communication from the Commission to the Council', *SEC* (91) 1744 final

European Commission (1992a) 'Completion of the internal market in electricity and gas', *COM* (91) 548

European Commission (1992b) 'A Community strategy to limit carbon dioxide emissions and to improve energy efficiency, Communication from the Commission', *COM* (92) 246 final

European Commission (1992c) 'Proposal for a Council Directive introducing a tax on carbon dioxide emissions and energy', *COM* (92) 226 final

European Commission (1992d) 'Proposal for a Council Directive to limit carbon dioxide emissions by improving energy efficiency (SAVE programme)', *COM* (92) 182 final

European Commission (1993a) 'Council Directive 93/76/EEC to limit carbon dioxide emissions by improving energy efficiency', *Official Journal of the EC*, Vol 237, pp28–30

European Commission (1993b) 'Specific actions for greater penetration for renewable energy sources – ALTENER', *Official Journal of the EC*, Vol L235, 18th September 1993

European Commission (1994) 'Assessment of the expected CO_2 emissions from the Community in the year 2000', *SEC* 94 (122)

European Commission (1995a) 'Review of the Fifth Environmental Action Programme', *COM* (95) 624

European Commission (1995b) 'Proposal for a Council Directive to introduce rational planning techniques in the electricity and gas distribution sectors', *COM* (95) 369 final

European Commission (1995c) 'Commission working paper on the EU climate change strategy: a set of options', *SEC* 95 (288) final

European Commission (1996a) 'Directive 96/57/EC on energy efficiency requirements for household electric refrigerators, freezers and combinations thereof', *Official Journal of the EC*, Vol L236, 18th September 1996, pp36–43

European Commission (1996b) 'Second evaluation of national programmes under the monitoring mechanism of Community CO_2 and other greenhouse gas emissions', *COM* (96) 91

Eurostat (1996a) 'CO_2 emissions from fossil fuels 1985 to 1994', as quoted in *European Environment*, 30th April 1996, ppI.1–3

Eurostat (1996b) 'Aspects statistiques de l'économie énergétique en 1995', *Statistiques en Bref Énergie et Industrie*, no 22, pp5–8

Golub, J (1994) *The Pivotal Role of British Sovereignty in EC Environmental Policy*, Working Paper 94/17, Robert Schuman Centre, European University Institute, Florence

Golub, J (1996) *Sovereignty and Subsidiarity in EU Environmental Policy*. Robert Schuman Centre Working Paper 96/3, European University Institute, Florence

Grubb, M (1996) *Renewable Energy Strategies for Europe, Volume 1: Foundations and Context*, Earthscan, London

Grubb, M and Hope, C (1992) 'EC climate policy: where there's a will ...', *Energy Policy*, Vol 20, no 11, November 1992, pp1110–1114

Haigh, N (1996) 'Climate change policies and politics in the European Union', in O'Riordan, T and Jäger, J (eds) *Politics of Climate Change: A European Perspective*, Routledge, London

IEA (1995) *Energy Policies in IEA Countries*, IEA/OECD, Paris

Jachtenfuchs, M (1994) 'International policy-making as a learning process: the European Community and the greenhouse effect', PhD thesis, European University Institute, Florence

Klimabündnis (1993) *Klima – Lokal geschützt*, Raben Verlag, München.

Klimabündnis (1995) *Das Klimabündnis zum klima-Gipfel*, Klimabündnis, Frankfurt

Krämer, L (1995) *EC Treaty and Environmental Law*, Sweet and Maxwell, London

Macrory, R and Hession, M (1996) 'The European Community and climate change: the role of law and legal competence' in O'Riordan, T and Jäger, J (eds) *Politics of Climate Change: A European Perspective*, Routledge, London

McGowan, F (1991), 'Conflicting objectives in European energy policy', in Crouch, C and Marquand, D (1991) *The Politics of 1992*, Blackwell, London, pp121–137

Rotmans, J, Hulme, M and Downing, T E (1994) 'Climate change implications for Europe: an application of the ESCAPE model', *Global Environmental Change* Vol 4, no 2, pp97–124

SEA (1987) *Single European Act*, Office for Official Publications of the European Communities, Luxembourg

Sioshini, F (1996) 'DSM in transition: from mandates to markets', *Energy Policy*, Vol 24, no 4, pp283–284

Skjaerseth, J B (1994) 'The climate policy of the EC: too hot to handle?', *Journal of Common Market Studies*, Vol 32, no 1, pp25–45

Wynne, B (1993) 'Implementation of greenhouse gas reductions in the European Community: institutional and cultural factors', *Global Environmental Change*, Vol 3, no 1, pp101–128

Chapter 5 | # LEADERSHIP AND UNIFICATION: CLIMATE CHANGE POLICIES IN GERMANY

Michael Huber[1]

INTRODUCTION

Since the early 1980s, Germany has striven for leadership in environmental policy, both domestically and at the EU level. The country has, for example, been instrumental in fostering the EU directives on car emissions and on emissions from power stations. Germany declared its goal to become the policy leader in the field of climate change by issuing a comprehensive response programme in the areas of energy, transport and technological development policy, and setting itself the most ambitious target among all Member States of the EU, namely a 25 per cent reduction of CO_2 emissions by 2005, compared to 1990 levels. However, implementation of the policy programme has been affected negatively by two external factors. Firstly, the unification of the two German states in 1990 resulted in a shift in environmental policy focus, available resources and public attention. Secondly, the re-emergence of traditional, well known alliances and conflict lines over the nuclear issue has played a decisive role, resulting in a deadlock over energy policy.

Before unification environmental issues were considered a high priority in the former Federal Republic of Germany (FRG), according to public opinion polls, but this priority has declined considerably since the economic problems and costs related to the transition from a planned to a market economy have become evident. As a result, stagnation in policy developments regarding climate change has occurred since 1992. This does not

---

1 The author is grateful to Ute Collier for her contribution to the discussion of disagreements over nuclear power, a renewable energy boom, energy efficiency measures and local authorities as policy leaders.

imply that German climate policy has disappeared, rather that policy making shifted its focus towards the less demanding international and supranational levels. In this chapter an attempt is made to reconstruct the main elements of the emerging climate policy, indicate the problems and opportunities associated with the unification process and identify the main reasons for the observed implementation gap.

THE GERMAN EMISSION PROFILE

Germany is the EU's largest emitter of greenhouse gases. As Table 5.1 shows, CO_2 is by far the most important gas, with 1031 Mt emitted in 1990. Electricity generation accounts for the largest share of CO_2 emissions, followed by industry, the commercial sector and the transport sector. For CH_4, the largest share of emissions is associated with waste (landfill sites), while for N_2O, industrial processes are the largest culprit. The rest of the chapter focuses on CO_2 as the largest source of emissions.

Table 5.1. *Greenhouse Gas Emissions, Total and Share of Sectors*

Gas	Total	Share of sectors	
		Sector	Per cent
CO_2	1031 Mt	Power plants	35.6
		Industry	23.5
		Commercial	19.9
		Transport	18.8
		Non Energy	2.2
CH_4	6000 kt	Waste	36.7
		Agriculture	34.2
		Fossil Fuels	25
		Others	4.2
		Industry	45.5
N_2O	220 kt	Agriculture	34.1
		Power plants	5.9
		Transport	4.1
		Others	10.5
CFCs	37 kt		

Source: BMU, 1994

Germany accounts for about 5 per cent of global CO_2 emissions and 30 per cent of the EU total. In terms of tonnes per capita emissions, with 11.05 t per capita in 1993 Germany was also among the top emitters, ranking fifth in the EU (IEA, 1995). Quite clearly, emission developments in Germany are crucial to achieve the current EU emission stabilization target.

Unification has considerably worsened the German CO_2 balance. Of the 1990 emissions, nearly 30 per cent were emitted in the former German Democratic Republic (GDR) (subsequently referred to as the 'new *Länder*') which had per capita CO_2 emissions of 18.6 t, compared to 11.2 t per capita for the old *Länder*. This high level of emission was a function of three main factors:

- an industrial structure based on heavy industry;
- electricity generation based on inefficient lignite plants;
- great inefficiencies in industrial processes and domestic energy use.

However, at the same time, transport related emissions were low. In 1990, in the new *Länder* transport accounted for only 8.1 per cent of emissions, while in the old *Länder* it accounted for 22.5 per cent. However, this picture is changing fast, as industry in the new *Länder* has collapsed, energy efficiency investments are being made and the transport sector is growing rapidly. Emissions had already started declining before unification, with a reduction of 13.6 per cent between 1987 and 1990. Since then the reduction has been dramatic, as Figure 5.1 shows.

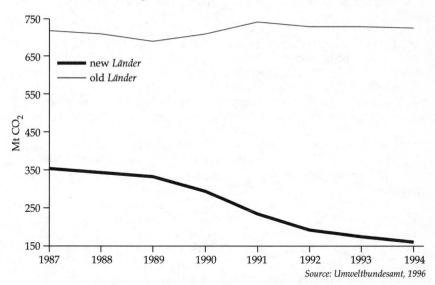

Source: *Umweltbundesamt, 1996*

Figure 5.1: CO_2 emission developments in Germany

Emissions in the new *Länder* decreased by more than half between the 1987 level (the highest in history) and 1994. This decrease is mainly due to the collapse of industrial production in the new *Länder*. Meanwhile, emissions in the old *Länder* have actually shown an increase since 1990.

Coal as a CO_2 intensive fuel plays an important role in the old *Länder*, but also in the new *Länder* where lignite still represents a large, although

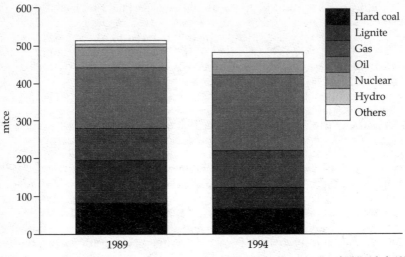

Source: Bundesministerium für Wirtschaft, 1995

Figure 5.2: Main fuels in German primary energy consumption

rapidly falling, proportion of primary energy consumption (44.5 per cent in 1994, down from 68.5 per cent in 1989). Lignite has the highest carbon content of all fossil fuels, so a change of fuels could lead to considerable emission reductions (see Figure 5.2).

Electricity generation in Germany is dominated by coal and lignite (56 per cent) and nuclear power (29 per cent). Final energy users are about 41 per cent for households (24.5 per cent) and small users (16.3 per cent); 31.6 per cent for industry and 27.6 per cent for transport. Final energy uses are 28.9 per cent for space heating, 25.2 per cent to generate process heat, 38.7 per cent for machines and transport, 5.2 per cent for water heating and light requires only 2 per cent.

The expected development of primary energy use will result in a growth in energy consumption in the old *Länder*, while in the new *Länder* a further decrease of energy consumption in the areas of production and domestic use is estimated. In the new *Länder* the only sector with increasing consumption is transport.

Overall, it looks increasingly unlikely that Germany will achieve its ambitious CO_2 reduction target for 2005. The most recent projection by the Prognos Institute for the Economics Ministry projects a reduction of only 10.5 per cent by 2005. A further drop by 3 per cent is expected by the year 2020.[2]

---

2 *European Energy Report*, No 450, 5th January 1996, p 15.

THE EVOLUTION OF CLIMATE CHANGE POLICY IN GERMANY

The Environmental Policy Background

Germany, or rather the former Federal Republic of Germany (FRG) before unification, has over the past 15 years or so acquired an image as one of the environmental policy leaders in Europe. Environmental policy emerged on the political agenda in 1969. In that year, the social democratic–liberal coalition came to power and the climate of reforms opened the stage for environmental issues, which had gained increasing public awareness and concern. The Liberal Party (FDP) saw an opportunity to acquire a better political profile by pursuing environmental policies. In 1971 the first environmental policy programme was adopted, based on the principles of cooperation, precaution and the polluter pays. At the institutional level, the Ministry for the Interior was given responsibility for environmental matters. However, by the mid 1970s environmental issues disappeared from the political agenda, although they were implicit in the conflict over nuclear power which ensued.

In the early 1980s, international issues such as acid rain and ozone layer depletion helped to re-establish an environmental agenda. In 1986 environmental concern surrounding the Chernobyl accident triggered the establishment of a separate Environment Ministry (Bundesministerium für Umwelt, BMU), since the Ministry of the Interior was perceived to be unable to manage the effects of such a disaster. Pehle (1995) observes that even though the first years of the BMU's operation were considered successful, a considerable implementation gap affected the proposed policies. While environmental issues remained high on the political agenda during the 1980s, since the unification between the FRG and the GDR in 1990 they have lost some of their salience, especially as the high economic costs of unification began to be felt. Nowadays, environmental policy measures have to be justified increasingly in terms of their cost neutrality, as the argument that environmental policies destroy jobs has gained importance. Concern about *Standort Deutschland* (ie the competitive position of the German economy) has become a central focus of political discussion, and is now an important influence on the German response to the climate change issue.

It is also important to consider the attitude to environmental issues in the former GDR. The dominant impression of the GDR as a country that had no environmental policies is misleading. The GDR had developed a comprehensive and well accepted body of environmental laws but their implementation and potential control was difficult. State owned firms were considered by the administration to be able to introduce waste reduction technologies and pollution control by themselves – with the result that most firms were unable to generate feasible solutions. In sectors such as agriculture the large cooperatives did not have the training to handle toxic substances. There was no public access to information about the state of the environment from 1974 as this was classified 'confidential'. After 1982 access to information on the environment was forbidden even to ministers and leading party officials, environmental problems being managed by the Minister of Economy personally. The results of such a 'non-policy' have been described as catastrophic and disastrous, with the development of many severe local problems related to toxic waste or high levels of air pollution (see Bachmann, 1993). Environmental concerns played an important role in the movement which eventually resulted in the fall of the Berlin Wall. With unification, the abatement of these very immediate environmental problems became a priority. However, more recently environmental issues have disappeared from the political stage, as issues of welfare and unemployment gained importance. This has affected particularly the more long term, less tangible issues such as climate change.

The Role of the First Enquete Commission

The climate issue gained public attention in Germany with a report on climate change by the *Deutsche Physikalische Gesellschaft* (German Physical Society) in late 1985. This triggered, together with international events and conferences, political interest in the issue. The first response by the Government was the establishment of an Inquiry Commission in 1987, as it had done previously for important matters such as the future role of nuclear power in the early 1980s. The Enquete Commission 'Protecting the Earth', consisting of Members of Parliament and scientists, commissioned a comprehensive research programme comprising 150 independent studies, and produced three reports between 1987 and 1992 (Enquete Commission, 1989; 1990; 1992) which provided an important basis for subsequent decisions by Government. Three scenarios for emission reductions of 30 per cent by 2005 were developed, using different assumptions about options such as energy efficiency, renewable energy sources and nuclear power and projecting reduction costs of between 2.7 billion DM and 9.4 billion DM (Enquete Commission, 1990).

In 1990, the German government reacted to the statement by the Enquete Commission that a 30 per cent reduction of CO_2 emissions by 2005 was feasible by issuing two decisions. The first decision in June 1990 announced a goal of a 25 per cent reduction of CO_2 by 2005, compared to the reference year of 1987 (BMU, 1990). The second decision in November 1990 simultaneously extended and weakened the first decision; it included the new *Länder* and stated that, due to the transition from a planned to a market economy, far reaching savings of CO_2 emissions could be obtained (BMU, 1991a). However, the numeric goal was no longer considered a strict target but a target towards which German policy was to be oriented. In December 1991 the German Parliament (*Bundestag*) accepted the target of the Enquete Commission to try to reduce CO_2 emissions by about 30 per cent by the year 2005 and 80 per cent by the year 2050, based on 1987 levels (BMU, 1992a). The *Bundestag* also decided a reform package for energy supply, transport and construction work. It is interesting to note that all the actors, ie government, members of the Enquete Commission, various ministries and the *Bundestag* have agreed on a common problem definition, saying that:

- climate change is an urgent problem,
- energy policy is the most appropriate policy area to manage the problem;
- a stringent target is required.

All members of the Enquete Commission agreed that the traditional 'muddling through' of German energy policy is insufficient to meet such a goal. In 1995, at the first FCCC conference of the parties, the Government strengthened the 2005 target by announcing 1990 as the base year. As CO_2 emissions in 1990 were 5 per cent lower than those of 1987, the target will be more difficult to achieve.

Elaborating the National Programme for Climate Protection

In 1991 a second Enquete Commission was established. Its task was to follow up the work of the first Commission and specify, parallel to an inter-ministerial working group (*Interministerielle Arbeitsgruppe*, IMA), possible policy paths, and to expand the time horizon from 2005 to 2020 and 2050. Also the changed situation resulting from unification called for a thorough analysis of the role of the new *Länder*. The aim was to find a feasible mix of instruments to reflect contextual factors such as the IEM, the relation between domestic and imported fuels and the economics of unification. While the work of the First Enquete Commission received an overwhelm-ingly positive reaction, the second Commission had to deal with more politicized issues and it has been suggested that its work has been domi-nated by controversial discussions, mistrust and frustration (Beuermann and Jäger, 1996). While there was quite a heated debate surrounding the issues of energy efficiency and savings, as well as of fuel switching, it was the conflict between nuclear and renewable energy proponents which repro-duced the polarization of the German energy debate of the early 1970s (Michaelis, 1995; see also pp73–74).

The actual elaboration of the German climate change programme was the task of an IMA, which was given the remit to elaborate a strategy for reducing CO_2 emissions by 25 per cent by 2005, with priority to be given to market based instruments (for a detailed discussion see Unfried, 1994). Five subworking groups were established, led by different ministries, and the IMA's final report was published in 1993; this provided the basis for the national programme, which was first published in 1993, and then, in a revised version, was submitted to the United Nations in 1994. The IMA started from the premise of the Third Report of the First Enquete Commission (1992) and drew up a comprehensive set of feasible measures.

The IMA estimated the technological potential of currently available, yet not fully developed, technologies which could have an impact on the reduc-tion of CO_2 emissions, as well as technologies that might offer potential for future medium and long terms reductions. The role of CHP, renewable ener-gies (wind, solar thermal and photovoltaics) and the scope for the expansion of nuclear energy were discussed. For CHP and renewable energies, a number of administrative and economic constraints were identified, in particular the current low level of energy prices. In the case of nuclear power, the IMA argued that public acceptance has to be improved before this technology could offer an alternative to current energy production.

In the final programme the IMA proposed 109 measures for the climate change programme, ranging from a CO_2 tax to specific regulations in the insulation sector (BMU, 1994). A number of the measures were already in existence and none were quantified in terms of expected CO_2 emission reductions, implying that the effectiveness of individual measures could not be verified. According to Beuermann and Jäger (1996), the majority of the measures implemented to date can be characterized as 'subsidies' promoting useful activities or investments aided by federal financial aid. Other measures affect only small groups (utilities, CFC producers) and the CO_2 reductions associated with them are incidental.

The CO_2 Tax and Voluntary Commitments from Industry

The most unpopular measure was the CO_2 tax, which to date has not been implemented and after years of polemics looks unlikely to ever come into being. The tax was promoted by many scientists and environmental organizations but was denounced by conservative political actors as unfeasible and expensive (Hey and Brendle, 1994). Industrial associations vehemently opposed such a tax, although several important industrial leaders admitted the necessity to open the debate on environmental taxation. The political tension around the CO_2 tax idea was partially due to its frequent interpretation as the first step towards a comprehensive green tax reform. In this context, it is interesting to observe how a substantial increase in petrol taxes was implemented in 1994. As its goal was to raise new revenues for unification, it was accepted without any opposition. In early 1995, the discussion about energy and CO_2 taxes re-emerged with the decision to abolish the coal support mechanism, the *Kohlepfennig* (see pp74–76), but reference to the environmental dimension blocked further debates. This episode exemplifies that it is not the real economic burden of taxes that is decisive for their political feasibility, but rather the reasons for their justification.

Some of the discussion focused on whether Germany should apply a tax unilaterally or if EU agreement should be found. When the EU carbon/energy tax began to look increasingly unlikely, calls from environmental groups and the opposition Social Democratic Party (SPD) for a unilateral tax intensified. In 1995 both the SPD and the FDP, the junior partner in the Bonn coalition, drew up plans for a 'climate tax', to be implemented as of January 1997. Industrial opposition remained strong and in spring 1996 the Government conceded to industry's calls to allow a voluntary agreement instead.

Already, in 1991 industry had offered to carry out voluntary measures to achieve emission reductions. It argued that a national target should be established between industry and government, and then it should be left to industry to decide how to achieve it. In 1995 industry[3] made a renewed offer to reduce its specific CO_2 emissions by up to 20 per cent (basis year 1987) by the year 2005 (BDI, 1995). Subsequently, an agreement was signed between the Government and 19 industrial and trade federations which accounted for nearly four fifths of industrial energy consumption. Industry has strengthened its commitment by agreeing to a definite reduction of 20 per cent by 2005 with 1990 as a basis year. In return, the Government has committed itself to refrain from imposing a carbon tax unilaterally and, in the case of an EU tax, to argue for exemptions for industry.[4]

The main problem of this agreement regards the notion of specific CO_2 emissions. It implies that each sector will reduce emissions on its own, that is regardless of the initial position of the sectors. In the appendix to the common declaration, single sectoral (or sometimes subsectoral) targets are listed, separate for the old and new *Länder*. Generally, the commitment relies almost entirely on reductions in the new *Länder*, which are happening in any case as a byproduct of industrial restructuring. It is not clear whether any new specific measures will be implemented to achieve emission reductions in the old *Länder* and the agreement is not very ambitious.

ENERGY MEASURES

Disagreements over Nuclear Power

In the discussion of German energy and climate policy, three energy options play a crucial role: nuclear power, renewables and coal. Conflict over nuclear energy, which currently contributes 11 per cent of Germany's energy requirements, has structured the energy debate in Germany since the beginning of the 1970s, and the climate issue was interpreted by the nuclear industry, as well as by the ruling Christian Democratic Union (CDU) party, as one justification for a resurgence of this form of energy production. Reduction scenarios outlined by the Enquete Commissions saw nuclear power as a main instrument in continuing a supply oriented energy policy while reducing CO_2 emissions at the same time. However, nuclear power has not regained public credibility, despite considerable support from the environment minister and the governmental coalition. The political support for nuclear power contrasts with the position taken by the energy supply industry. Here, problems with licensing procedures, political feasibility (especially in the *Länder* governed by the SPD and the Green Party) and large uncertainties regarding costs and safety have led to a certain reluctance by the energy supply industries to invest in nuclear plants. Furthermore, nuclear power continues to be strongly opposed by

---

3 The German car industry did not take part in this agreement. It had, through voluntary agreements, already reduced fuel use between 1978 and 1985 by 23 per cent, but was not willing to enter a new agreement before the EU-wide regulation of maximum emissions had been established.

4 *European Energy Report*, 12th April 1996, No 457, p 2; *Stromthemen*, No 5, May 1996, p 3.

the opposition party, SPD. The Second Enquete Commission has pinpointed these issues and the main obstacles for an energy consensus in Germany but has failed to suggest a way to overcome these constraints.

The well-researched polarization of public and political opinion (see eg, Kitschelt, 1980) was amplified in 1986 when the SPD decided to phase out nuclear power. This subsequently resulted in some SPD governed *Länder* delaying commissioning procedures for nuclear power, as in the case of the Mühlheim-Kährlich[5] plant in Hessen. The economic implications of these delays made the energy companies increasingly reluctant to invest in nuclear power. In December 1992 a letter to the German Chancellor by the Chairmen of the Boards of two of the largest energy companies, VEBA and RWE, with the political support of the Prime Minister of Lower Saxony, Gerhard Schröder (SPD), pointed out the need for a broad, societal agreement on energy policy to reduce the uncertainties of future developments. Subsequently, the Government decided to initiate the so-called energy consensus talks.

In December 1992 the Ministers of the Environment and Economy were delegated to represent the Government in the discussions, which in principle were to focus on the future of energy policy in general, including the role of coal and renewable energies. In practice, they centred upon the role of nuclear power. However, after two and a half years of confrontational discussions, the talks were abandoned in the summer of 1995. No consensus was reached and several groups, including Non-Governmental Organizations (NGOs), the Green Party and trade unions, left the discussion process early. Environmental groups and the Green party refused to even consider a future for nuclear power, while the SPD and the trade unions requested that any discussion about nuclear power must be linked to the continuing support of the German coal industry. However, the SPD's commitment to coal subsidies and the *Kohlepfennig* indicates its problems with an energy future beyond the traditional paths. As most coal production takes place in SPD governed *Länder*, a change in coal consumption would imply job losses which would have a negative impact on regional political activities.

The energy consensus talks did highlight the structural problems of German energy policy and demonstrated a certain decoupling of the needs of energy policy and party politics. However, they also showed that the climate change issue, at least with the current level of concern, is unlikely to have a major impact on the future of nuclear power. For the foreseeable future, it is thus improbable that any new nuclear plant will be built in Germany.

REDUCING EMISSIONS FROM FOSSIL FUEL ELECTRICITY GENERATION

The German electricity sector has officially committed itself to reduce CO_2 emissions by 15 per cent by the year 2015. Considering that there is little

---

5 The electricity industry has in recent years increasingly argued for the commissioning of this plant on CO_2 emission reduction grounds.

likelihood of an expansion in nuclear generation, the achievement of this target will have to depend on fuel switching to gas, using renewable energies and improved generation efficiency. In principle, a switch to gas, as is currently occurring in the UK (see Chapter 6) could be envisaged, in particular as the heavy government support of German hard coal has been under continuous attack over recent years, both from industry and from the European Commission. This led to a challenge of the main support mechanisms, the *Kohlepfennig* (which in 1995 amounted to an 8.5 per cent surcharge on electricity bills), in front of the constitutional court, with the result that it had to be abolished by 1996. Coal subsidies now have to be financed directly by the federal budget but it is likely that this support will be reduced over the coming years.

However, as coal is the only domestic fuel in Germany, there are a number of problems in employing a radical substitution strategy to achieve the CO_2 reduction target. Loske and Hennicke (1993) have argued that a choice of fuels should not be based on current market prices alone, but should consider issues of dependency, long term developments, technology transfer and social justice. Firstly, the import costs of oil and gas in 1992 amounted to 39.1 billion DM and, in macroeconomic terms, it is undesirable that this burden be increased. Secondly, closing down the few remaining coal mines would hit two industrial areas in Germany which are already weakened by the decline of the steel industry. Thirdly, it can also be argued that there is a considerable market for the export of coal technologies, as countries such as India and China begin to exploit their resources.

The only way a coal support policy can be made compatible with CO_2 emission reduction objectives is by developing new efficient coal technologies and by exploiting the potential for CHP. There is already considerable investment in new coal technologies, including a coal gasification plant built by RWE. RWE, the biggest energy utility, is also the parent company of Rheinbraun, the largest lignite company in the old *Länder*, and has thus a particular interest in ensuring the future of coal fired generation. CHP is already widespread in Germany, with many towns having extensive district heating grids. According to COGEN (1995) there is almost as much CHP in Germany as in the rest of the EU put together. CHP accounts for around 15 per cent of total installed generation capacity and just under 10 per cent of electricity generation. As is shown in more detail later (pp80–82), local authorities, in conjunction with municipal energy companies, have focused their energy policies in recent years on the expansion of CHP, combined with district heating (DH), partially as an oil replacement strategy and, partially for environmental reasons. Bewag, the Berlin municipal utility, operates the largest DH grid in Western Europe and it estimates that this has reduced CO_2 emissions in Berlin by 8 per cent.[6] In recent years, there has been a particular expansion in mini- and micro-CHP plants. However, there are now fears that deregulation in the electricity market will make CHP–DH less competitive as it cannot necessarily compete with other fuels.

---

6 *Stromthemen*, No 7, July 1995, p 3.

Furthermore, it requires long term, stable investment planning, which is difficult to achieve in a competitive climate.

Despite a general commitment to the use of domestic coal, there has been considerable investment in gas fired plant and it is expected that gas fired plants will account for 61 per cent of all new capacity to be built until 2020.[7] A number of gas fired CCGT plants have already been commissioned or are under construction in the new *Länder*, generally with a CHP element. In other cases, old lignite plants have been renovated and some new lignite CHP plants built. Compared to inefficient old lignite plants, considerable CO_2 reductions are taking place. Between 1990 and 1995 CO_2 emissions by power stations had already decreased by 7 per cent and further decreases are likely until 2000.[8]

A Renewable Energy Boom?

Renewable energies have been the long time favourites of the Greens and have also received support from the opposition party SPD as an alternative to nuclear power. Germany has currently a very small contribution (only 0.5 per cent) from renewable energies to the overall energy requirements. They contribute just under 5 per cent of electricity generation, most of which comes from hydropower (3.9 per cent). While well established, the potential for large hydro plants has been largely exhausted. Although, generally, Germany is not considered one of the most favourably positioned countries for renewable energy sources, there is still reasonable potential for small hydro, wind, biomass and solar power developments. In the 1990s renewable energies experienced something of a boom. From 1992 to 1994, electricity generation from renewables increased by 11 per cent.

While various government R&D programmes [eg the 250 MW wind energy programme and the 2250 roof photovoltaic (PV) programme] have been important, the crucial factor in this development was the 1991 Energy Feed Law (EFL) (*Stromeinspeisungsgesetz*). This law was adopted in response to the strong environmental and antinuclear lobby, which requested a level playing field for renewable energies. The law requires energy companies to purchase electricity produced by renewable energies and sets down minimum reimbursement fees, far above 'avoided cost' levels.[9] Effectively this means that the extra costs of renewable energy (DM 135 million in 1995) have to be paid for by the energy companies, with particularly high costs for those companies that operate in areas with a high potential for renewable energies (eg Schleswag in the North, where there has been a large expansion in wind power). These extra costs cannot necessarily be passed on to the consumer, as electricity prices are regulated by the *Länder* authorities who have not been generally sympathetic to increased prices. Another problem is that the EFL also benefits the owners of existing hydropower plants which already operate on a commercial basis and as such need no subsidy.

---

7 Report by Prognos Institute, op cit Footnote 2.
8 *European Energy Report*, No 458, 16th April 1996, p 16.
9 The energy companies claim that reimbursement levels are at between 30 and 70 per cent above the long term avoided costs for plant construction and fuel purchase (*Stromthemen*, No 5, May 1996, p 5).

Not surprisingly, there has been forceful opposition to the EFL from the large energy companies. The Southern German utility Badenwerke decided to suspend payments to a hydropower operator and was subsequently taken to court. At the time of writing, the decision is with the constitutional court and there have been discussions about finding a mechanism to allow the spreading of the financial burden across all utilities. The association of electricity companies (VDEW) has been calling for the abolition of the EFL, but this seems unlikely considering the large popular support for renewables. Meanwhile, some energy companies have their own separate support programmes for renewables. RWE, for example, has just launched a DM 20 million solar programme, providing a subsidy of DM 2000 to customers for the installation of solar panels, PV systems or heat pumps.

Wind power has been the biggest beneficiary of the EFL, helped by additional government grants. It has been estimated that the technical potential of wind power in Germany is as high as 117 TWh, 20 per cent of gross electricity generation.[10] Current production is just over one tenth of this figure and at the end of 1995 wind capacity stood at 1127 MW, an astonishing 78.3 per cent increase over the previous year's figure.[11] Much of the coastal potential has already been exhausted and attention is now focusing on inland sites. German turbine manufacturers are putting a considerable effort into the export market. However, as in the UK, public opposition to wind turbines has been increasing.

In general, despite some positive developments, the contribution of renewable energies is likely to remain small, at least in the medium term. The future of the EFL is rather uncertain, and it is not clear whether the government would be prepared to offer funds to compensate for an eventual abolition of the law.

Energy Efficiency Measures

According to the National Programme, energy efficiency improvements on both supply and end use sides have to be a priority area of climate change policy. A number of measures are listed but, generally, federal measures in this area are not very far reaching. During the 1980s there was a federal programme to provide grants and subsidies for a number of energy efficiency investments (in particular building insulation), but they expired in the early 1990s (Collier, 1994a). Grants have been made available in the new *Länder* and these have certainly been important in ensuring improvements in their notoriously badly insulated housing stock. Meanwhile, insulation standards for new buildings were tightened in 1995 and are expected to cut heat requirements by 30 per cent. However, according to Beuermann and Jäger (1996), these standards were watered down in the course of negotiations between the federal level and *Länder* administrations.

---

10 *European Energy Report*, July 1995, supplement, p 12.
11 *Renewable Energy Report*, No 15, 24th May 1996, p 6.

A considerable level of activity is occurring at local and regional levels, as well as in the large energy companies. All companies are involved in activities such as energy advice provision, as well as subsidies for certain efficient appliances. RWE, the largest company, has been running its customer energy efficiency programme *KesS* (*Kunden-Energiespar-Service*) since 1992 and has made available over 100 million DM, providing a subsidy of DM 100 for the purchase of particularly energy efficient appliances. Since 1995 the programme has been restricted to refrigerators.[12] Furthermore, RWE is running energy efficiency programmes for industrial and local authority consumers. A number of municipal companies have been very active in this area, as discussed in more detail on pp80–82.

A major problem facing the initiatives for energy efficiency is that, as elsewhere, low energy prices provide a substantial disincentive to investment. The abolition of the *Kohlepfennig* is resulting in further reductions in electricity prices. One positive development has been the revision of the electricity tariff regulation (*Bundestarifordnung Elektrizität*). This allows electricity companies to change their tariff structures and reduce fixed charges. Many companies now have linear and time variable tariffs, which give a greater energy saving incentive to some customers, in particular smaller consumers.

THE TRANSPORT DIMENSION

Germany probably has the best road system in the whole of Europe and one of the highest levels of car ownership, with BMW and Mercedes cars being one of the symbols of the German economic miracle. The automobile industry plays an important role in the German economy and any attempt to question the role of the motor car is bound to arouse opposition from a strong industrial lobby. Germany is also famous for its absence of speed limits on the motorways, an issue of importance for CO_2 emissions as, invariably, higher speeds mean higher petrol consumption and hence higher emissions. However, despite the strong dominance of private transport, Germany actually has a good public transport system, with many local authorities running extremely efficient transport systems. The federal railway system has also received high levels of government subsidies.

When the Government decided to privatize the federal railway company (*Bundesbahn*) in 1994, it argued that an efficient, effective and well developed network of public transport would be able to compete with cars and road based freight transport. However, it refused to free the railways of their substantial debts before privatization. It is interesting to note that the members of the privatization review body were tightly linked to car producers (Volkswagen), oil companies (Shell), banks (Dresdner Bank), insurance companies and the association of German industry (BDI) (Paulitz, 1994). The initial results of privatization have been somewhat mixed. While huge

---

12 *Stromthemen*, No 4, April 1995, p 7.

investments have been made to improve the city transport of the railways by Intercity Express lines, smaller lines have been closed down and goods transport has been moved from the tracks to lorries owned by the railways.

Meanwhile, the demand for road based transport is growing unabated. The growth pattern in the new *Länder* is particularly problematic. In the former GDR, private transport was considerably less developed than in the former FRG but the 'motorization' of the population has been speedy, to the detriment of CO_2 emissions. Projections show that even if there was a comprehensive programme to mitigate CO_2 emissions, an increase in emissions of about 87 per cent in private transport and 67 per cent in goods can be expected. Without interventions these figures are are expected to be even higher, as Table 5.2 shows.

Table 5.2. *Expected Transport Developments in the new* Länder

		1988	2005 Trend (% change)	2005 Reduction (% change)
Private transport	primary energy (Mtce)	137	229 (+67)	213 (+55)
	CO_2 emissions (Mt)	12800	33000 (+158)	23900 (+87)
Freight transport	primary energy (Mtce)	77	153 (+99)	131 (+70)
	CO_2 emissions (Mt)	5920	11400 (+93)	9870 (+67)
Total	primary energy (Mtce)	258	617 (+139)	463 (+79)
	CO_2 emissions (Mt)	19200	44700 (+133)	34100 (+78)

Source: BMU, 1991b

Despite the high level of these projected figures, the transport dimension of climate change policy has received minor public and political attention. The transport subgroup of the IMA, together with two independent advisory bodies, TH Aachen (Aachen Technical University) and the consultancy firm PROGNOS AG, reviewed the proposals of the Enquete Commission and developed a set of feasibility scenarios. A list of 22 CO_2 reduction measures formed the basis of three scenarios. Scenario A proposed nine measures which focus on the infrastructure of public transport and provide (weak) incentives to changes in individual behaviour. None of these measures were considered by the BMU to be effective. Scenario P used 13 measures which constitute a more compact policy package, while the most comprehensive Scenario R utilized all 22 measures. A reference scenario (2005 Trend) represented the expected worst case scenario in terms of CO_2 emissions. The impact on CO_2 emissions of these scenarios is summarized in Table 5.3.

Table 5.3. *Projected Developments in the Transport Sector*

	Billion km travelled	Change compared to 1987 (in %)
1987	406.3	± 0
2005 Trend	495.6	+ 22
Scenario A	432.1	+ 6.4
Scenario P	324.3	- 20.2
Scenario R	239.5	- 41.2

Source: BMU, 1992b

These figures were criticized by the Federal Environmental Office (Umweltbundesamt, UBA) for not being realistic. UBA felt that to achieve emission reductions, additional measures were necessary such as a set of fiscal measures to increase the prices of cars by 10 per cent and the costs of car use. Additionally, a halt to the construction of new motorways would be required. In addition, UBA recommended that revenues raised from higher personal transport costs should be used to improve public transport, while strict controls of goods transport, such as speed limits, higher charges for the use of motorways and strict technical standards, should allow for better competition between road and rail freight transport. However, in the current political climate none of these measures appear feasible and the transport dimension of climate change policy is effectively nonexistent, at least at the federal level.

LOCAL AUTHORITIES: THE REAL POLICY LEADERS?

A number of local authorities are pushing forward with their own policies, in both the energy and transport sectors. Germany has some of the best political conditions for local action on climate change. Firstly, the right to self government for municipalities is enshrined in the *Grundgesetz* (the constitution). Secondly, municipalities are the majority shareholders in the *Stadtwerke*, the local utility companies for energy, transport, water and waste services. They can thus exert a major influence over the activities in these sectors. Furthermore, environmental awareness is particularly well developed in Germany, with the Green Party being well represented on most local councils. Local authorities also boast long established environment departments.

Most local climate change plans have evolved out of the so-called local energy concepts, which many local authorities have been engaged in since the 1980s. Their aims were to integrate security of supply (especially to reduce oil dependency) and environmental protection objectives (Collier, 1994b). Many authorities also considered them a symbolic rejection of nuclear power, a stance which continues, notwithstanding the more recent concerns over climate change. While the policies differ between local authorities, they all have some common characteristics:

- Plans and programmes have been drawn up jointly between local authorities, local councillors and local public utility companies (*Stadtwerke*);
- preference for CHP power plants;
- integrated heat planning with priority areas for district heating;
- energy efficiency programmes with subsidies;
- some investment in renewables (albeit limited);
- large subsidies for public transport.

As already mentioned in Chapter 4, German local authorities were behind the establishment of the Climate Alliance, which has been joined by over 200 towns and cities in Germany. Of these, 16 are also members of the ICLEI 'Cities for Climate Protection' campaign. The local authorities of Heidelberg, Saarbrücken and Leipzig have shown themselves to be particularly proactive and high profile in their climate protection activities. The successes and problems of the strategies of these four cities are discussed briefly here.

Saarbrücken has won a number of prizes for its plans to realize a sustainable energy system, including a United Nations prize at the Rio summit. The '*Zukunftskonzept Energie*' (energy concept of the future) was originally conceived in 1980, with a major focus on the heating sector aimed at the replacement of oil fired boilers through the expansion of district heating and gas (Stadtwerke Saarbrücken, 1991). Initially, the rationale was to improve security of supply and to reduce acid emissions, although climate change has become an increasingly important focus of the programme. Particularly interesting in climate change terms is the commitment to utilize coal as a local resource (both for security of supply and job protection reasons), but to reduce its emissions through efficient CHP generation. A subsidy scheme for PVs also exists, aiming at the installation of 1000 kW (0.5 per cent of overall municipal electricity demand) by 2000. Furthermore, there are various energy efficiency information activities and grant and loan schemes. Through these activities, CO_2 emissions were reduced by 15 per cent between 1980 and 1990 (Stadt Saarbrücken, 1993).

A climate protection programme was produced by the town council in 1993, aimed at a further reduction of CO_2 emissions of 25 per cent by 2005. However, the achievement of this target is proving a challenge. A team of consultants has drawn up a reduction strategy, specifying investments of 800 million DM, 300 million of which would be required by the *Stadtwerke*. The strategy requires substantial energy savings which would at the same time result in a decrease in the *Stadtwerke*'s turnover by 8.3 per cent (Masuhr, 1994). Unless this can be compensated through price increases or innovative financing frameworks, implementation of the plan is unlikely. However, neither the federal nor the state government appear favourable to such changes. As far as emissions from the transport sector are concerned, plans for a tram network could result in significant reductions but are highly dependent on state and federal subsidies, the availability of which is not assured.

Heidelberg became one of the environmental 'leaders' among German local authorities with the election in 1991 to the mayor's office of Beate Weber, previously chairperson of the European Parliament's environment committee. Heidelberg was a prime mover behind ICLEI's Heidelberg declaration. After commissioning a consultancy report on how to achieve emission reductions, a working group was instituted, composed of all four political parties in the local council, the *Stadtwerke* and council officers. Based on its deliberations, in spring 1993 a catalogue of priority measures was agreed by the town council. A cornerstone of the strategy is the reduction of emissions from municipal buildings and facilities. Although only accounting for 2.9 per cent of total CO_2 emissions in the town, the importance of setting an example is seen as crucial, apart from the fact that considerable financial savings also result, currently amounting to 1 million DM per annum (Stadt Heidelberg, 1994). To increase connections to the district heating grid has been a policy for some time and includes the dedication of DH priority areas. The reduction of emissions from the transport sector is proving the greatest challenge and focuses on high subsidies for local transport services and improvements to existing services. Furthermore, a policy of decentralization of council offices has cut down car journeys. Increasing public awareness is considered an important part of the climate change strategy, so a climate change advice centre has been established in the council's environment office.

Leipzig exemplifies the rather different situation in the new *Länder*. There, the previously state-owned energy companies were to be divided up between the large old *Länder* energy companies according to the *Stromvertrag* (Electricity Act) of 1991. However, this was contested before the constitutional court by 164 authorities and a compromise agreement awarded them the right to form their own *Stadtwerke* (Collier, 1994a). Some of these, such as Leipzig, have now drawn up environmentally focused energy policies, albeit with a main emphasis on the reduction of air pollution. However, large CO_2 emission reductions will result as well. Both the heating and electricity sectors were almost exclusively based on inefficient brown coal plants which are now gradually being replaced. The heating sector was dominated by district heating although this was generally supplied by heat only plans, transported in uninsulated, overground pipelines. Priorities are now the renovation of these networks, the construction of efficient CHP plants and a range of energy efficiency measures, for which there are some federal subsidies. Leipzig is currently building a 172 MWe gas fired CHP plant with plans for a second. This, together with other measures and a substantial reduction in energy use through industrial restructuring, is expected to result in CO_2 emission reductions of over two thirds by 2000. Such reduction figures are obviously quite unique in the EU, but could probably be achieved elsewhere in Eastern Europe.

CONCLUDING REMARKS

Germany has striven to become a leader in climate change policy by setting itself the most ambitious CO_2 reduction target in the EU. Emission reductions have occurred in recent years but mainly as a result of the collapse of industrial production in the new *Länder*. In the former FRG, emissions have actually increased since 1990. However, unification has been as much a curse as a blessing for German climate change policy. Unification substantially affected the political climate in Germany, in particular when its large economic costs became apparent. First of all, this meant that allocations in the federal budget changed. As far as environmental policy was concerned, the allocation of ever scarcer state resources meant some difficult decisions. It has been estimated that, overall, DM 211 billion of investments are required in the period up to 2000 to clean up immediate environmental problems such as air pollution, water pollution and contaminated land (Bachmann, 1993). Hence, not surprisingly, priority was given to these issues over the more long term measures related to climate change. Secondly, the increased tax burden imposed on citizens and businesses meant that acceptance of costly climate change measures, such as a carbon tax, has not been forthcoming. Karger et al (1993) found that only a minority would accept measures that economically burden their everyday activities.

Nevertheless, there are a number of positive developments in the new *Länder* as regards CO_2 emissions. Efficiencies in both electricity generation and end use have been improving over recent years as a large scale effort is under way to replace inefficient lignite power stations and to renovate buildings to relatively high energy conservation standards. Efforts are also being made to renovate the extensive district heating grids that exist in many town and cities in the new *Länder*, which are now being linked to new CHP plants, often gas fired. Government grants and subsidies have been made available to support these developments. Meanwhile, in the former FRG, there is much commitment to climate change policies at the local level.

However, overall, it has to be noted that despite a number of favourable conditions, climate change policies in German cities are encountering some difficulties. Over recent years, local authority budget constraints have set limits to investments in DSM or renewables, as the *Stadtwerke* have come under increased pressure to produce high returns. In some towns, parts of the *Stadtwerke* have been sold for short term revenue, thus reducing the councils' abilities to influence their investments and activities. Furthermore, falling energy prices, especially through the abolition of the *Kohlepfennig*, give a disincentive to consumers as far as energy efficiency investments are concerned. Finally, liberalization in the energy sector, for which there is increasing pressure both from industry and the European Commission, could have negative effects in that it would undermine the position of the *Stadtwerke*.

In Germany, huge endeavour has been put into the study and analysis of possible policy options to achieve emission reductions. A number of new ideas have been introduced into the debate, such as the ecological tax reform. However, despite considerable political support, these ideas have made little headway, mainly because of industrial opposition. Furthermore, previous discussions about the future of nuclear power have resurfaced and the energy consensus dialogue is seen as an attempt at reaching a solution. In this context, the Government's attempt to use the climate change issue as a justification for its support for nuclear power has essentially failed.

Overall, the future prospects for emission reductions in Germany are somewhat mixed. Reductions by the new *Länder* are likely to reach a plateau soon. While industry will be restructured and much more efficient, growth will invariably mean emission increases. Meanwhile, measures to reduce emissions by the old *Länder* are modest and the voluntary agreement with industry is unlikely to result in any new reduction measures. Furthermore, little has been done to address the fast emission growth from the transport sector. This issue is intricately linked with the German car industry and any radical measures to reduce the role of the motor car will face substantial opposition.

REFERENCES

Bachmann, G (1993) *Environmental Policy in Transition: The Need for a New Political Approach to Environmental Clean Up in the Former GDR*, Working Paper, European University Institute, Florence

Beuermann, C and Jäger, J (1996) 'Climate change politics in Germany: how long will any double dividend last?' in O'Riordan, T and Jäger, J (eds) *Politics of Climate Change: a European Perspective*, Routledge, London

BDI (1995) *Erklärung der Deutschen Wirtschaft zur Klimavorsorge*, Bund deutscher Industrie, Köln, 10th March 1995

BMU (1990) *Beschluß der Bundesregierung zur Reduzierung der CO₂ Emissionen in der Bundesrepublik Deutschland bis zum Jahre 2005*, BMU, Bonn

BMU (1991a) *Beschluß der Bundesregierung vom 7. November 1990 zur Reduzierung der CO₂-Emissionen in der Bundesrepublik Deutschland bis zum Jahre 2005 auf der Grundlage des ersten Berichtes der IMA CO₂-Reduktion*, BMU, Bonn

BMU (1991b) *Vergleichende Analyse der in den Berichten der Enquete Kommission 'Vorsorge zum Schutz der Erdatmosphäre' und in den Beschlüssen der Bundesregierung ausgewiesenen CO₂ Minderungspotentiale und Maßnahmen*, BMU, Bonn

BMU (1992a) *Beschluß der Bundesregierung vom 11. Dezember 1991: Verminderung der energiebedingten CO₂-Emissionen in der Bundesrepublik Deutschland auf der Grundlage des zweiten Berichtes der IMA CO₂-Reduktion*, BMU, Bonn

BMU (1992b) *Ermittlung und Bewertung von CO₂ Minderungspotentialen in den neuen Bundesländern der Bundesrepublik Deutschland*, BMU, Bonn

BMU (1994) *Environmental Policy. Climate Protection in Germany. First Report of the Government of the Federal Republic of Germany Pursuant to the United Nations Framework Convention on Climate Change*, BMU, Bonn

Bundesministerium für Wirtschaft (1995) *Energiedaten 95*, Referat für Öffentlichkeitsarbeit, Bonn

COGEN (1995) *The Barriers to Combined Heat and Power in Europe*, Cogen Europe, Brussels

Collier, U (1994a) *Energy and Environment in the EU*, Avebury, Aldershot

Collier, U (1994b) 'Local energy concepts in Germany: an environmental alternative to liberalisation?', *Energy and Environment*, Vol 5, no 4, pp305–326

Enquete Commission (1989) 'Eine internationale Herausforderung; Zwischenbericht der Enquete-Kommission des 11. Deutschen Bundestages "Vorsorge zum Schutz der Erdatmosphäre"', *Zur Sache*, Bonn

Enquete Commission (1990) 'Eine Bestandsaufnahme mit Vorschlägen zu einer neuen Energiepolitik; Dritter Bericht der Enquete-Kommission des 11. Deutschen Bundestages', *Zur Sache*, 1/2, Bonn

Enquete Commission (1992) *Klimaänderung gefährdet globale Entwicklung; Erster Bericht der Enquete-Kommission "Schutz der Erdatmosphäre" des 12. Deutschen Bundestages*, C F Müller, Bonn

Hey, C and Brendle, U (1994) *Umweltverbände und EG: Strategien, politische Kulturen und Organisationsformen*, Westdeutscher Verlag, Opladen

IEA (1995) *Energy Policies in IEA Member States*, IEA/OECD, Paris

Karger, C, Schütz, H and Wiedemann, P M (1993) 'Zwischen Engagement und Ablehnung: Bewertung von Klimaschutzmaßnahmen in der deutschen Bevölkerung', *Zeitschrift für Umweltpolitik und Umweltrecht*, Vol 16, no 2, pp201–215

Kitschelt, H (1980) *Kernenergie. Arena eines politischen Konflikts*, Campus, Frankfurt/M

Loske, R and Hennicke, P (1993) 'Klimaschutz und Kohlepolitik, Überlegungen zu einem strukturellen Dilemma deutscher Energiepolitik', *Wuppertal Papier* No 5, September, Wuppertal Institutes für Klima, Umwelt, Energie, Wuppertal

Masuhr, P (1994) 'Das Saarbrücker Umsetzungprogramm zum Klimaschutz: betriebs- und regionalwirtschaftliche Betrachtungen', presented at the conference *Saarbrücker Energiestudie 2005*, 18th April 1994, Saarbrücken

Michaelis, H (1995) 'Bilanz der Arbeit der Enquete Kommisison Schutz der Erdathmosphäre. Leitlinien der Klimapolitik', *Energiewirtschaftliche Tagesfragen*, No 1/2

Paulitz, H (1994) *Manager der Klimakatastrophe. Die Deutsche Bank und ihre Energie und Verkehrspolitik*, Verlag Die Werkstatt, Göttingen

Pehle, H (1995) 'Umwelt- und Umweltaußenpolitik in Deutschland', paper presented at the workshop on New Nordic Member States and the Impact on EC Environmental Policy, Sandbjerg

Stadt Heidelberg (1994) *Handlungsorientiertes kommunales Konzept zur Reduktion von klimarelevanten Spurengasen für die Stadt Heidelberg*, Stadt Heidelberg

Stadt Saarbrücken (1993) *Klimaschutzprogramm der Stadt Saarbrücken*, Stadt Saarbrücken

Stadtwerke Saarbrücken (1991) *Das Saarbrücker Zukunftskonzept Energie*, Stadtwerke Saarbrücken

Umweltbundesamt (1996) 'Emissionen nach Emittentengruppen', obtained through personal communication

Unfried, M (1994) 'Regierungspolitik gegen Klimakatastrophe. Die deutschen CO_2 Minderungsbeschlüsse 1990/91 und die Schwierigkeiten einer querschnittorientierten Umweltpolitik', Erlangen, unpublished thesis, University of Erlangen

Chapter 6 | 'WINDFALL' EMISSION REDUCTIONS IN THE UK

Ute Collier

INTRODUCTION

During the 1980s, the UK was frequently labelled the 'dirty man' of Europe (Rose, 1990), stemming primarily from its reluctance to reduce high levels of sulphur emissions from power stations, which were reputedly contributing to the destruction of Scandinavian forests and lakes. Unlike in many other North European countries, environmental issues were low on the British political agenda during this period. A change in attitude from late 1988 onwards is usually attributed to a speech by the then Prime Minister, Margaret Thatcher, on the importance of environmental protection. Simultaneously, public awareness of environmental issues was growing and with it came calls for a more proactive environmental policy.

Climate change, together with ozone depletion, was one of the major issues that benefited from this new attention. The Government initially set a target for the return of CO_2 emissions to 1990 levels by 2005. This target was brought forward to 2000 in spring 1992. Subsequently, the Government prided itself in being the first country to submit its CO_2 reduction strategy under the FCCC. The most recent emission projections indicate possible CO_2 emission cuts of between 4.4 and 7 per cent and the Government now tends to portray the UK as being at the forefront of the international effort to combat climate change.

The aim of this chapter is to show that far from suddenly becoming a leader in climate change policy, the UK is still dragging its feet in many important policy areas and environmental concerns are often compromised by other policy priorities. The projected emission reductions are a matter of accident rather than design (the 'windfall' referred to in the title), resulting from the 'dash for gas' subsequent to electricity privatization, and are likely to be short lived. Meanwhile, policies in the energy efficiency field, where there remains much scope for emission reductions, are very weak. Greater policy commitment can be found in some local authorities but their scope for action is limited.

EMISSION TRENDS IN GREENHOUSE GASES

In the UK in 1990, CO_2 emissions accounted for 87.3 per cent of overall greenhouse gas emissions, followed by CH_4 (8.1 per cent) and N_2O (4.4 per cent) (HM Government, 1994). The UK is the second largest CO_2 emitter in the EU and in terms of per capita emissions and emissions per unit of GDP it is in sixth place. With these emission characteristics, the UK has no claim to any special dispensation, unlike some of the other Member States.

Recent years have shown a clear fall in CO_2 emissions, as Table 6.1 shows, despite a growth of GDP.

Table 6.1. *UK CO_2 Emissions (1990–1995)*

	1990	1991	1992	1993	1994	1995
Total CO_2 emissions (Mt)	158	159	155	151	149	148
Reasons for change:	*Increase or decrease in emissions (%)*					
Temperature differences		+5	+2	+2	+2	+1
Changes in GDP		–3	–4	–1	+6	+10
Amount of energy used per unit GDP		+1	+2	+1	–4	–7
Use of lower carbon fuels		–2	–3	–9	–12	–14
Overall change		+2	–3	–7	–9	–10

Source: Department of the Environment, 1996

According to the Department of the Environment (1996), in 1992 the drop in emissions was mainly due to the economic recession of that year. Subsequently, emission reductions have been primarily a result of the shift to natural gas in electricity generation (see below). Furthermore, the contribution to electricity generation from nuclear power has been higher than expected, mainly due to an increased performance of the Advanced Gas Cooled Reactors (AGRs), which had previously suffered from various performance problems. Emissions have seen a reduction despite a significant growth in GDP.

According to the most recent government projections (*Energy Paper 65*, Department of Trade and Industry, 1995), emission reductions of between 4.4 and 7 per cent between 1990 and 2000 can be expected. These reductions will be almost entirely due to fuel switching in electricity generation. As Figure 6.1 shows, *Energy Paper 65* projects a considerable shift between the various fuels used in electricity generation. While in 1990 coal accounted for

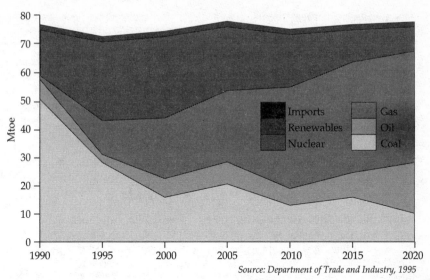

Source: Department of Trade and Industry, 1995

Figure 6.1: Fuel use in UK electricity generation[1]

68 per cent of all fuel use in electricity generation, this is currently decreasing steadily and may be as low as 20 per cent by 2000. Meanwhile, the share of natural gas is set to increase from zero to 34 per cent. Nuclear power will temporarily increase its share to 34 per cent in 2000, before declining to 3.5 per cent in 2020. However, the increased share of nuclear power is somewhat uncertain as the performances of individual plants have been very variable in the past.

Due to the varying growth expectations in other sectors, the relative importance of different sectors as an emission source is expected to shift. While in 1990 power stations accounted for the largest proportion of CO_2 emissions (34 per cent), they will be overtaken by transport by the year 2000. From then onwards, emissions from transport and from power stations are expected to continue in more or less equal proportions. Emissions from the domestic sector, industry and agriculture are expected to stay fairly constant, while those from the service sector are likely to increase slightly.

In 1990, CH_4 emissions amounted to around 5 Mt, with the largest proportion of the emissions (39 per cent) coming from landfill sites, followed by agriculture (32 per cent) and coal mining (16 per cent). The Government expects that CH_4 emissions will be reduced to about 4.4 Mt in 2000. Most of this reduction will be due to the continuing decline of coal production in the UK. Furthermore, some emission reductions from landfill sites are also likely, as landfill gas plants are receiving support under a renewable energy scheme (see pp97–99).

---

1 This figure is based on the Central GDP Growth–Low Fuel Prices scenario of *Energy Paper 65*. This is essentially a business-as-usual scenario, assuming a GDP growth of 2.35 per cent per annum and low fuel prices (eg $15 per barrel of oil).

As concerns other greenhouse gases, the Government expects a decline in all emissions. N_2O emissions are expected to fall substantially from around 0.11 Mt in 1990 to 0.03 Mt in 2000, mainly as a result of control strategies in industry. A downward trend can also be observed in NO_x emissions but it is less marked from 2.8 Mt to between 1.8 Mt and 2.2 Mt by 2000. However, it is clear that NO_x control from motor vehicles in the UK is progressing only very slowly, as there is no programme to retrofit catalytic converters. The same applies to carbon monoxide emissions.

THE UK CLIMATE CHANGE PROGRAMME

Climate Change and the Political Agenda

When the climate change concern first started to hit the headlines in 1988–1989, the Government initially did little more than refer to the role of British scientists in developing the scientific understanding of the issue. At the time, a number of newspapers reported nightmare scenarios involving the flooding of large parts of the British Isles, an image which much of the population still adheres to. However, Parry and Duncan (1995) argue that current defensive systems (which cover 33 per cent of British coastlines) should be able to cope with the consequences of published best guess estimates of future sea level rise. Also, in terms of agricultural production, the UK (especially the uplands) could benefit from climate change as cropping patterns shift north. However, the many uncertainties involved in modelling the costs and benefits of climate change mean that it would be a considerable risk to rely on the assumption that the UK would have a net benefit from climate change.

One of the first government activities was the publication by the now defunct Department of Energy[2] in January 1990 of a report on energy related greenhouse gas emissions and measures to ameliorate them (Department of Energy, 1990). Environmental groups and opposition parties continued to call for action but none was taken by the Government until Margaret Thatcher, the then prime minister, announced, quite unexpectedly at the opening of a new climate research centre in May 1990, a target for the stabilization of CO_2 emissions by 2005. At subsequent EU level discussions, the UK government refused to fall in line with other Member States and adopt a 2000 target date. This refusal has been attributed to the desire to proceed smoothly with electricity privatization and not to alienate private motorists before the 1992 election.[3] As is shown later, privatization has indeed been an overriding priority in energy and transport policy, with a mixed outcome in CO_2 emission terms.

2 The Department of Energy was abolished after the 1992 elections. The Government argued that as most public energy companies had been privatized, its role in the energy sector was much reduced and hence there was no longer any need for a separate government department. Most of the Department's functions were moved to the Department of Trade and Industry, but the Energy Efficiency Office was relocated to the Department of the Environment.
3 *The Guardian*, 26th May 1990, pp 1, 2.

The climate change issue coincided with a general growth in environmental awareness in the UK and the strategy to implement the CO_2 target was first spelt out in the Government's first ever policy document on the environment, the 1990 White Paper entitled *This Common Inheritance*. The White Paper claimed that the Government was taking 'major initiatives' on energy efficiency (HM Government, 1990) but, as later sections show, these have in fact to date been rather limited. Furthermore, it was claimed electricity privatization would to lead to a 'very substantial' reduction in CO_2 emissions. Indeed, as the scale of the 'dash for gas' became apparent, the Government brought forward its CO_2 stabilization target to 2000 in April 1992, finally falling in line with the EU target. Since the favourable projections of *Energy Paper 65* emerged, the Government has prided itself for being at the forefront of international efforts to combat climate change (Department of the Environment, 1996).

The UK Climate Change Programme

The two most significant policy measures to emerge subsequent to the Environment White Paper have been the imposition of VAT on domestic fuel and the establishment of the Energy Saving Trust (EST, see below), two measures which then became the backbone of the UK CCP submitted under the FCCC. The CCP was elaborated by the Department of the Environment and in the process a consultation document was produced and various discussions with a variety of interest groups took place. The final product was published in January 1994 (HM Government, 1994) and contained a strong emphasis on voluntary measures, termed the 'partnership' approach by the Government. The plans were for an overall saving of 10 Mt of CO_2 by the year 2000. This was to be achieved through various measures, as listed in Table 6.2.

Table 6.2. *CCP 1994 Measures for CO_2 Reductions*

Measures	Emission reductions (MtC)
Energy consumption in the home (including VAT on domestic fuel, EST, ecolabelling, building regulations, appliance standards)	4
Energy consumption by business (including EEO programmes, building regulations, office machinery standards)	2.5
Energy consumption in the public sector (through target setting)	1
Transport (increased road fuel duties)	2.5
Total	10

Source: HM Government, 1994

The division of emission reductions shown in Table 6.2 is somewhat confusing in that it lists the CO_2 reductions to be achieved by sector, with no specific allocation to each measure. The CO_2 reductions achieved through fuel substitution on the electricity generation side (gas, renewables, CHP) are supposed to be integrated in these figures. It was stated separately in the CCP that VAT at 17.5 per cent was supposed to account for 1.5 Mt of CO_2 reductions and the activities of the EST for 2.5 MtC.

In the following sections the main measures of the CCP are discussed in more detail. As is shown, VAT and the EST have encountered major difficulties and the Energy Efficiency Office (EEO) programmes are rather small scale. These problems have been acknowledged by *Energy Paper 65* and the 1996 progress report, which has revised the figures of the contribution of the CCP. According to this, the CCP is now expected to yield reductions of only 7.5 Mt of CO_2 (Department of the Environment, 1996), although considering other developments, such as falling prices (see pp172–174), even this scaled down figure seems overoptimistic. Furthermore, the Government does not appear to have taken into account the effect of the lower carbon density of electricity supply in its calculations of reductions due to energy efficiency measures.[4]

CHANGES IN ENERGY RELATED EMISSIONS: RELYING ON THE 'DASH FOR GAS'

Energy Efficiency and the Energy Saving Trust Debacle

According to the initial CCP, before the windfall emission reductions from the electricity sector became apparent, energy efficiency was to be the cornerstone of the UK's climate change strategy, with a particular reliance on the newly established EST. The EST was launched in May 1992 by the Government as a non-profit making company, in response to the many criticisms of the Government's energy efficiency policy (Collier, 1995). The Trust was set up as a joint partnership between the Government, British Gas, the Regional Electricity Companies[5] (RECs), Scottish Power and Scottish Hydro Electric. The aim was to develop, propose and manage programmes to promote energy efficiency so as to stimulate markets. The Trust started with a budget of under £5 million in 1993–1994 and was supposed to reach an investment profile of £400 million by the year 2000. Most of this funding was to come from surcharges on gas tariff and electricity franchise customers. By 1997–1998 it was expected that finance from these two areas would reach £150 million each (EST, 1994). However, it is now clear that this was an extremely optimistic assumption.

The main problem has been the reluctance of the regulators of the gas and electricity industry, Office for Electricity Regulation (OFFER) and Office

---

4 *ENDS Report*, No 251, December 1995, p 21.
5 The RECs are responsible for the distribution and supply of electricity to final customers.

of Gas Regulation (OFGAS), to sanction this level of expenditure.[6] Initially, when the EST was set up, the then director of OFGAS, James McKinnon, introduced a small 'e-factor' to raise £2 million for EST pilot projects, which was mainly used to provide subsidies for the installation of condensing boilers. However, his successor, Clare Spottiswoode, decided in 1994 to reverse this decision by claiming the e-factor constituted a tax she was not entitled to raise (O'Riordan and Rowbotham, 1996). Hence, the EST has not received any further funding from British Gas. Furthermore, OFFER's direc-tor, Stephen Littlechild, set the contribution from the electricity sector at a much lower level, amounting to only £100 million in total over 4 years (£1 per customer per year). The RECs have to run EST approved DSM programmes which are linked to Standards of Performance. These require companies to achieve a certain amount of demand reduction, amounting to around 2.3 per cent of current supply, which is not very ambitious. Most RECs have been rather reluctant to become involved in energy efficiency promotion, as they cannot make any profits from this activity. The electric-ity sector contribution is to cease with the introduction of full competition in the electricity sector in 1998.

To avert the funding dilemma for the EST, in May 1995 the Government committed itself to provide a core funding of £25 million per annum ('until gas and electricity markets are fully liberalized'), a rather modest sum, which was then cut to £15 million in the November 1995 budget. In its 1996–1997 programme, the EST is focusing on the following measures:

∎ subsidies for condensing boilers;
∎ rebates for central heating control units;
∎ support for local authority energy efficiency strategies.[7]

While these measures are potentially important, the lack of sufficient funding limits the number of subsidies that can be given. A main dilemma is that since privatization the government has limited influence over the regulatory system for the energy companies. The primary legislation on the duties of the regulators is not specific enough as regards energy efficiency or general environmental duties. The Government had an opportunity to change this unsatisfactory situation, at least for the gas sector, when it intro-duced a new gas bill to parliament in 1995, with the aim of introducing competition. The Round Table on Sustainable Development[8] and environ-mental organizations lobbied for the inclusion of an environmental duty on the regulator, but this was rejected by the Government. The Government's justification for this has been that in a competitive market, suppliers would use energy efficiency as a means of taking business from others,[9] which is

---

6 The regulators are quasi-autonomous; while the primary legislation (the electricity and gas bills) outlines their responsibilities, they have quite wide powers of discretion. The directors are government appointed but are subsequently independent in their actions.
7 *ENDS Report*, No 255, April 1996, p 26.
8 The Round Table was set up by the Government to advise on the implementation of the sustainable development strategy and has 30 members from business, environmental organi-zations, local government and other bodies.
9 *The Guardian*, 9th March 1995.

also the argument used by OFFER as far as the electricity market is concerned.[10] However, it is far from certain that energy efficiency will be used as a selling point in the competitive market. A recent report by the Round Table on Sustainable Development (1996) suggests that competition based on prices rather than energy services is more likely, at least initially. They have identified a clear need for a regulatory framework which encourages companies to develop energy services. As yet, no details have emerged on the exact structure of the 1998 domestic sector liberalization, but the experience to date does not really point towards a willingness to intervene heavily. However, a change of government in 1997 might have positive implications in this area, as the Labour Party has signalled that it will put more emphasis on energy efficiency (Labour Party, 1994).

Apart from the EST, the EEO, attached to the Department of the Environment, also runs a number of schemes. The largest of these in budget terms is the Home Energy Efficiency Scheme (HEES) which was launched in 1990[11] and aims to increase the uptake of basic insulation measures in low income households by means of grants and advice. This scheme receives a large part of the EEO's budget, amounting to £100 million in 1994–1995. However, the budget was cut back by £31 million for 1996–1997, which throws further doubts as to the Government's commitment to the promotion of energy efficiency. Also, according to Boardman (1993), most of this has been spent on draught-proofing measures with a limited life time and hence has probably not been the most cost effective use of the money. Furthermore, low income housing at which the HEES is targeted is often quite good in energy efficiency terms, as many improvements were made by local authorities, who previously owned most social housing.

Apart from a number of smaller programmes that provide advice on energy management in business, the EEO runs the Making a Corporate Commitment Campaign (MCCC), which seeks top management commitment to energy efficiency from private and public sector organizations. It aims to increase the awareness about the links between environmental protection, energy efficiency and profitability. On joining the campaign, a chairman or chief executive of a company signs a declaration stating that the company is committed to responsible energy management and will promote energy efficiency throughout its operations.

The Government does not negotiate specific targets with the companies, so that they can choose these themselves. Accountability to shareholders is supposed to ensure the achievement of the performance targets.[12] To date, more than 1800 companies and public sector organizations have joined the MCCC but it is not clear what the accumulative achievements of the campaign will be, as no attempt has been made to calculate the potential total savings. Without any clear targets, except at the company level, it will be very difficult to monitor the progress of the

---

10 Interview, OFFER, March 1996.
11 But a similar scheme had previously been operated by the Department of Employment, so effectively no new resources were made available.
12 Interview, EEO, March 1996.

campaign and its contribution to the national CO_2 target. An EEO survey of 700 of the campaign signatories has revealed that only 75 per cent have set performance targets and only 70 per cent have an energy policy.[13] Furthermore, there are doubts whether the MCCC will achieve any additional savings beyond what the companies were going to do anyway.

VAT on Fuel: the UK's Answer to the Carbon/Energy Tax

The CCP clearly stresses the importance of pricing signals to domestic consumers as a key element. The Government's pricing strategy was to hinge on two measures: the introduction of VAT on domestic fuels and a continuous increase of excise duties on petrol. The VAT issue was very contentious from the beginning. Like a number of other essentials (food, books, children's clothing), domestic gas and electricity were zero rated for VAT purposes. VAT was to be introduced in two stages, 8 per cent by April 1994 and the full rate of 17.5 per cent by April 1995. There is little doubt that concerns about budget deficits rather than environmental issues were behind the decision to impose VAT. However, the issue was subsequently presented as an environmental measure, and an important component of the CO_2 reduction strategy. It also served as a means to oppose the proposals for the EU carbon/energy tax, as it enabled the Government to claim that it was already imposing a carbon tax.

The VAT plans have essentially fallen flat. Firstly, the issue was immediately hijacked by the Opposition, stressing the regressive nature of the measure. The Government managed to introduce the first 8 per cent of VAT but a vote in the House of Commons against the Government resulted in it having to abandon plans to raise this to the full VAT rate of 17.5 per cent. In the CCP progress report, the Government scaled down the expected emission savings due to VAT from 1.5 Mt CO_2 to 0.4 Mt CO_2. However, even this figure appears rather optimistic, as energy prices are on a downward trend. Firstly, both the gas and electricity regulators announced that they are clamping down on excessive profits made by British Gas and the RECs and hence have effected stricter price controls. Secondly, the Government announced in May 1995 that it would privatize the nuclear industry which would be accompanied by an 8 per cent cut in electricity bills. Consequently, over the next few years there will be cuts in domestic gas and electricity bills which will more than compensate for the imposition of VAT at 8 per cent. Incentives to invest in energy efficiency are thus being eroded continuously.

'Windfall' Emission Reductions in the Electricity Sector

As already mentioned, the main reason for CO_2 emission reductions occurring in the UK is a switch from coal to gas in electricity generation. Before privatization, British Coal had a guaranteed market for a large part of its output through a joint understanding with the Central Electricity Generating Board (CEGB) dating from 1979. The newly privatized genera-

---

13 *ENDS Report*, No 251, December 1995, p 21.

tors have been keen to switch to imported coal and gas as soon as possible for cost, as well as diversity, considerations and in order to comply with legislation to reduce emissions of SO_2 and NO_x. The Government's main concern has been the promotion of competition, believing that in a competitive market the generators could not be forced to burn large amounts of expensive coal (Collier, 1994, 1995).

Natural gas is partially set to take over coal's share in electricity generation, as National Power and PowerGen, RECs and independent generators have all been investing in CCGTs. Initially, new CCGTs were also attractive because, depending on gas prices, they can offer a cost effective way for the generators to meet their obligations to reduce SO_2 and NO_x emissions under the large combustion plant directive, avoiding the expensive retro-fitting of coal fired plants with flue gas desulphurization (FGD). More importantly, they offer independent generators the most economic method of entry into the system. As far as this concerns the RECs,[14] it allows them to reduce their dependence on the two main generators (House of Commons Energy Committee, 1992a). By June 1995, 19 CCGT plants were either operational or under commissioning or construction, taking the CCGT capacity to nearly 12,000 MW by 1997.[15]

While, on the one hand, there are CO_2 benefits in the move to gas fired capacity, the creation of overcapacity reduces the overall macroeconomic incentive to invest in energy efficiency as, in economic terms, generation companies will gain the most return on their investment by using the new plant to the full and finding a maximum market for the electricity generated. Furthermore, the CCGTs are all large scale plants with little use of waste heat or cogeneration. Little attention has been given to potential heat markets when deciding on power station siting. In Germany, in the new *Länder*, some new CCGT capacity is being located in town centres and linked to district heating grids. In the UK, there is a lack of such infrastructure and there are no incentives to install such grids. It is clear that the market alone will not accomplish the better net efficiencies that can be achieved through a greater use of CHP.

Nevertheless, privatization and liberalization have to some extent improved the situation for CHP. Industrial companies have been interested in CHP, as it offers a way of avoiding the payment of the fossil fuel levy (FFL, see below), if 50+ per cent of the electricity consumed was used on the premises. However, this incentive will cease with the abolition of the FFL and in addition there are a number of obstacles to the further development of CHP. Apart from the £2 million residential CHP grant system under the EST, which expired in 1996, there have been no special grants or tariff arrangements for CHP, and reimbursement prices are among the lowest in Europe (Cogen Europe, 1995). Another problem has been rising gas prices, combined with a price cap on electricity sold through the pool system. Also, the RECs have at times been obstructive to granting connections to the grid,

14 Before privatization, the old area boards were not allowed to generate electricity. The 1989 Electricity Act enables the RECs to generate up to 15 per cent of their supply requirements themselves but this limit is not being enforced.
15 Figures according to Ilex Associates, as reported in *Power in Europe*, No 201, 16th June, 1995, p 8.

as CHP systems undermine their sales.[16] The White Paper on the Environment set a target for 4000 MW_e of CHP by 2000, which was increased to 5000 MW_e by the CCP. At the end of 1994 there were 1167 CHP sites in the UK, with a total electricity capacity of 3141 MW, amounting to just over 4 per cent of total plant capacity. While the target for 2000 may be achieved, CHP capacity is not expected to greatly increase its share in overall generation capacity.

Renewable Energy: New Impetus

Although energy efficiency has so far benefited little under the regulatory system set up with privatization, renewable energies have received somewhat of a boost. As with fossil fuel resources, the UK is better endowed with renewable sources than most other EU Member States. The House of Commons Energy Committee (1992b) stressed that renewable sources of energy are a major national resource. So far, very little of the UK's renewable potential has been harnessed, which is at least partially due to the lack of any substantial government support. For example, during the early 1980s the Government all but abandoned support for wave power R&D.[17]

However, changes came with the growing concern about environmental issues. The Government set a target of 1000 MW of new renewable electricity generating capacity by 2000 in the 1990 Environment White Paper, which was subsequently increased to 1500 MW in 1993. The cornerstone of the renewables strategy is the Fossil Fuel Levy (FFL), a levy of around 10 per cent imposed on customers in England and Wales. According to the 1989 Electricity Act, the Secretary of State may make orders requiring each REC to secure the availability of a certain amount of electricity generated from non-fossil sources. This so-called Non-Fossil Fuel Obligation (NFFO) was originally established to protect the nuclear industry post privatization. After much lobbying by organizations that supported renewable energy, a separate 600 MW quota was established for renewables, to be filled in stages by 2000. By 1996, this quota had been long exceeded through four separate orders under the NFFO. The first three tranches contracted 1251 MW declared net capacity (DNC). For the fourth tranche almost 900 bids totalling 8400 MW DNC were submitted and it is expected that 400 to 500 MW of these will be approved by March 1997.[18]

Scotland and Northern Ireland were initially exempt from the NFFO. For Scotland, this decision was based on the already existing overcapacity in the electricity sector, mainly due to nuclear plants, notwithstanding that it has been estimated that 55 per cent of the UK wind energy resource is located in Scotland.[19] Public pressure brought a change in policy and the

16 Interview, Cogen Europe, March 1996.
17 This decision was based on an assessment of the economics of wave power for which it was later widely acknowledged that some gross miscalculations had been made (*The Guardian*, 16th February 1990, p 16).
18 *ENDS Report*, No 254, March 1996, p 13.
19 Although it has to be acknowledged that at the same time Scotland presents only 10% of the electricity market (with most of the demand in Southern England), and transmission losses would make any large scale expansion of wind power in Scotland uneconomic.

first Scottish Renewables Order (SRO) in 1994 offered contracts for 76 MW of capacity, aiming to achieve a final figure of 30–40 MW. However, progress has been slow and public opposition to wind developments has been making itself felt.[20] Two renewables orders have also been made in Northern Ireland.

Overall, 1343.27 MW DNC have been contracted under the different renewables orders so far. As Figure 6.2 shows, a large proportion of the capacity contracted is in the waste-to-energy area (incineration, landfill and sewage gas), the 'renewability' of which is somewhat debatable. Nevertheless, many of the schemes (in particular the landfill and sewage gas ones) are beneficial in greenhouse gas emission terms, as they contribute to reductions in CH_4 emissions. More of concern is the fact that only just over a quarter of the projects contracted are operational to date. A number of projects had to be abandoned because of planning objections, generally based on public opposition, especially in the waste incineration area, where only 6 out of 34 schemes have been commissioned.[21] A number of the 131 wind projects have also encountered problems. In other cases, there have been various delays. This situation may well result in a shortfall to the declared 1500 MW target if many of the schemes continue to encounter problems.

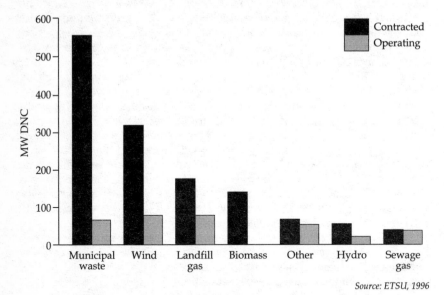

Source: ETSU, 1996

Figure 6.2: Renewable energy capacity under the Renewables Obligations (1996)

---

20 *Safe Energy Journal*, No 108, March–May 1996, p 21.
21 Personal communication, ETSU commercialization unit, June 1996.

Apart from support through the NFFO, renewables also benefit from a government R&D programme. In 1994–1995 this had a budget of £19.78 million, down from £25.6 million for the previous year. The budget is to be reduced over a 10 year horizon as the technologies 'move towards the market' (Department of Trade and Industry, 1994). This appears somewhat short sighted as a number of technologies, such as wave power, are likely to need support over a much longer term. As Elliott (1994) has pointed out, the NFFO has not proven to be well suited to some of the more radical, longer term renewable options. These often do not just require operating subsidies but capital grants, low interest loans, loan guarantees and other forms of capital underwriting to kick start the projects. Hence, while windpower, landfill and sewage gas, as well as waste incineration, have benefited greatly from government support, other, more long-term options such as wave power, tidal power or PVs have been almost totally neglected.

Overall, the introduction of the Renewables Obligation has been an important step towards a greater role for renewable energies in UK energy supply, although it is not yet certain whether the target of 1500 MW of renewable energy by 2000 will be achieved. Furthermore, this figure is small compared to the overall UK generating capacity and there has been too much emphasis on costs and not enough on environmental impacts, diversity and longer term options. Generally, liberalization is likely to present a double edged sword for renewable energies. On the one hand, there are some signs that full deregulation of the electricity market may benefit renewables, since the sale of 'green electricity' might find a market niche, as is already happening in the Netherlands and in Sweden.[22] Yet, at the same time, with falling electricity prices the commercial viability of renewables will be challenged further and will continue to require a support system of some kind. The future growth of renewable energies in the UK is thus far from assured.

Nuclear Power: A Victim of Privatization

Meanwhile, nuclear power quite clearly has no future in the UK. When the Government opted to privatize the electricity industry, the high costs of generation from this source emerged and it had little choice but to keep the nuclear power stations in state ownership and to cancel the construction programme for the pressurized water reactors (PWRs), with the exception of Sizewell B. Furthermore, it was decided that nuclear power in England and Wales[23] would be supported through the aforementioned FFL, which amounts to a major subsidy. However, the EU Commission ruled that the subsidy (except the renewables portion) had to be abolished by 1998, which made the future viability of nuclear power look uncertain.

Subsequently, Nuclear Electric greatly improved its performance during the early 1990s and lobbied heavily to be privatized and to be allowed to build another PWR at Sizewell. Part of its strategy was to use arguments about CO_2 reduction potentials but this had little impact. In May

---

22 *ENDS Report*, No 254, March 1996, p 27
23 Electricity generated from the two Scottish nuclear plants was exempt from the levy, as their generation costs were lower.

1995 the Government announced its plans to privatize part of the nuclear industry. These consisted of the privatization of Nuclear Electric's five AGRs and the Sizewell B PWR as well as Scottish Nuclear's two AGRs. The Magnox reactors, which are approaching the end of their working life, remained in the public sector and were transferred to British Nuclear Fuels, operators of the Sellafield nuclear reprocessing plant. However, the sale in June 1996 was far from successful, with 12.3 per cent of the shares not finding a buyer and the rest trading at below offer price on the stock market. The new company running the privatized plants, British Energy, has announced that it will not build any further nuclear reactors. Instead, it is planning to move into gas fired generation. As old plants are being retired, nuclear capacity will thus decrease considerably over the next 20 years, which will obviously have negative effects on CO_2 emissions. As the discussion has been totally dominated by economic arguments, this has not been a consideration. However, in terms of wider sustainability concerns, the demise of nuclear power has to be welcomed.

DEVELOPMENTS IN TRANSPORT POLICY

Developments in transport policy do to some extent mirror those in energy policy. Public transport provision has become increasingly influenced by the Government's preoccupation with the privatization and deregulation of previously publicly owned services. Until very recently, environmental issues had very little influence on the Government's thinking on transport policy. The transport aspect of the CCP itself relied almost entirely on an increase in the rate of road fuel duties by at least 5 per cent in real terms each year. Additionally, an initial three pence per litre (around 10 per cent) increase was imposed in 1993, with further increases in December 1994, after the Government's defeat on VAT, and in the November 1995 budget. However, falling petrol prices have cushioned the duty increases so it is not certain how much effect they will have.

The modal split between private and public transport is even more polarized in the UK than in most other EU countries. In 1991, the private car had an 88 per cent share in passenger km, compared to a combined 12 per cent for rail and buses. Of all freight 65 per cent was transported by road and only 8 per cent by rail. This percentage is currently declining even further. A further 28 per cent of freight was transported by water (RCEP, 1995). A forecast made by the Department of Transport in 1989 predicted an increase of between 83 per cent and 142 per cent (depending on assumed fuel prices and economic growth) in the level of road traffic by 2025 (Department of Transport, 1989). The Government's response to these forecasts was to draw up a major programme of public funding for new trunk roads and road improvements, while public transport continued to suffer from cuts.

The lack of an effective and environmentally sound transport policy was highlighted in 1995 by a report by the RCEP which received widespread media coverage. At the same time, a number of new road projects saw an unprecedented amount of public opposition and, according to Owens (1995), the established approach to transport policy was coming under sustained attack. The government responded with a Green Paper on transport in 1996, the first major policy document on the issue for nearly two decades. To some extent the Green Paper (Department of Transport, 1996) does indicate a rethink, acknowledging that greater account needs to be taken of the environmental impacts of transport. However, the Paper contains no quantified targets for future transport CO_2 emissions for or the fuel efficiency of vehicles, which the RCEP report had called for.

A key issue in view of the negative modal split is public transport provisions, which have been under constant criticism from environmental groups, mainly due to the perceived underinvestment compared to other European countries. Investment in the railways, for example, has been criticized for being only one quarter that of France or the Netherlands.[24] As already mentioned, transport policy has been affected by the Government's preoccupation with privatization and the introduction of competition. The first action in line with this policy in the transport sector was the deregulation of bus services in 1986. The negative environmental effect this is having in terms of CO_2 and other vehicle exhaust emissions, as well as other regressive effects of increased and uncoordinated public transport provision, are strongly criticized in the 1995 RCEP report.

Currently, the rail service is being privatized. It is not yet clear what the exact implications of this will be, but it is generally expected that it will involve line closures and higher prices. A report by the House of Commons Transport Committee (1995) suggested that the cost to the taxpayer would be an extra £600 million a year to fund current levels of services. A tight regulatory system is being established, which does involve the Office of Passenger Rail Franchising setting minimum standards of service. However, these do permit often substantial cuts in current services, which private operators are likely to adopt in many cases.

Some local authorities are planning and investing in new public transport provisions in the form of light rail transit systems. However, to date the schemes are small in number mainly due to the high capital costs and budget constraints. On the positive side, in coming years the balance between government funds for public transport and road investments will be redressed somewhat, according to the latest budget plans. Government expenditure on public transport will rise from £1.4 billion in 1995–1996 to a peak of £1.9 billion in 1997–1998. At the same time, the road budget will fall from £2.2 billion to £1.8 billion.[25] However, this is unlikely to be sufficient to stem the fast growth of CO_2 emissions from the transport sector. Meanwhile, the Green Paper passes the responsibility for achieving targets for the reduction of traffic growth to local authorities.

---

24 *The Independent*, 1st February 1995.
25 *ENDS Report*, No 255, April 1996, p 21.

CLIMATE CHANGE AND LOCAL GOVERNMENT

Local authorities in the UK have become increasingly active in drawing up their own environmental policies, especially as a means of implementing a local Agenda 21. According to Agyeman and Evans (1994), they are currently the leading actors and agencies in environmental policy and practice in Britain. While climate change is not the immediate focus of these policies, it is being tackled indirectly through measures in the transport and energy efficiency fields. Some local authorities (most notably Newcastle and Leicester City Councils) have drawn up specific energy strategies, although there is no legal requirement to do so.

In 1994, the environmental pressure group Friends of the Earth (FoE) started the 'climate resolution' campaign, attempting to encourage local authorities to sign up to a resolution to achieve by 2005 a 30 per cent reduction of the 1990 level of CO_2 emissions related to energy and transport use in the authorities' geographical areas. With the signature, local authorities undertake to develop within 12 months a detailed and comprehensive strategy to achieve the target (Friends of the Earth, 1994). To date 15 authorities have signed up to this resolution. Signature is of course entirely voluntary, and FoE sees the value of the resolution mainly in terms of persuading local authorities to think about CO_2 reductions and to take measures within their powers, rather than stick strictly to exact figures. Nine local authorities (Birmingham, Burnley, Eastleigh, Exeter, Leicester, Manchester, Sheffield, Strathclyde and Swale Borough) have also signed up to ICLEI's Cities for Climate Protection campaign.

Among the most active local councils in the UK are Cardiff, Leicester and Newcastle. Leicester was designated during the 1980s by the Department of Energy as one of the UK's lead cities for the development of CHP combined with district heating. A feasibility study concluded that CHP was indeed viable in Leicester. Subsequently, a company was formed to develop a CHP scheme but uncertainties surrounding the privatization of the electricity sector prevented the project from going ahead. Instead, the city council has installed a number of smaller gas fired CHP units. In 1995 a new energy strategy was drawn up, with some support from DG XVII. The strategy aims at identifying measures which will lead to a 50 per cent reduction in CO_2 emission (1990 level) by 2025 (Leicester City Council, 1995). To begin implementing this target, the Council has drawn up a separate Home Energy Strategy and Business Energy Strategy, consisting mainly of the provision of advice. Furthermore, high standards of energy efficiency for its new build properties have been adopted. However, the Council sees a large scale CHP system as essential in moving to a more sustainable energy system, yet under current market conditions this is not economically viable.[26]

Similar constraints have been encountered in Newcastle, which in 1992 prepared an 'Energy and Urban Environment Strategy', with funding from DG XVII. This suggested that CO_2 emissions related to energy use could be

---

26 Interview, Leicester City Council, March 1996.

reduced by 35 per cent by the year 2000 and 45 per cent by the year 2010 (1990 level) through a variety of measures, including a CCGT–CHP scheme, traffic restraint measures, radical improvements in energy efficiency and energy from waste schemes (Newcastle City Council, 1992). A slightly more modest target of a 30 per cent reduction by the year 2000 was subsequently adopted in the 1995 Unitary Development Plan. However, it does now seem unlikely that this target will be achieved, as much of it was relying on the CCGT–CHP scheme. This was to be a 150 MW plant to be built as a joint venture which would have produced 83 per cent of the city's electricity needs. The scheme relied on a number of institutional customers, including one of the universities, for their heat demand. However, in 1995 the university pulled out, so the scheme will only be able to proceed if a new customer can be found as a replacement.[27] So far, the city only has one district heating scheme in conjunction with a waste reclamation plant, which produces fuel pellets.

Cardiff is interesting in that it owns one of the few municipally run bus companies, thus being better placed for strategic transport planning and to influence emission reductions than most other councils. However, competition from other operators is continuously increasing, thus undermining the Council's efforts. The City and the County Council have together produced an Integrated Transport Strategy, which has as one of its main aims a change in the modal split from 60 per cent private motor vehicle and 40 per cent public transport trips into the city centre, to a 50:50 split (South Glamorgan County Council, 1994). However, the City Council foresees difficulties in achieving the strategy's goals because of continuously increasing levels of traffic into the city, in particular from former coal mining areas. Declining jobs have increased commuter traffic, mostly by car, into Cardiff substantially,[28] a trend which is difficult to influence.

Overall, a number of UK authorities are surprisingly active, but they face substantial obstacles in implementing their climate change strategies, ranging from budget constraints to lack of responsibility in crucial areas. One positive development is the Home Energy Conservation Act, which was passed by the House of Commons in March 1995,[29] and which for the first time gives UK local authorities specific responsibilities in the energy area. This requires local authorities to draw up strategic energy conservation plans for all residential property, both public and privately owned. The plans, to be submitted by November 1996, are to detail the practicable cost effective measures necessary to achieve 'significant' energy savings (subsequently defined in guidance notes as at least 30 per cent) and to specify the resulting CO_2 reductions.[30] While this legislation will provide much needed data on the energy efficiency of UK housing stock, it does not provide any extra finance for the actual carrying out of the plans and may thus have little effect. Also, local authorities are somewhat uncertain as to how exactly

---

27 Interview, Newcastle City Council, March 1995.
28 Interview, Cardiff City Council, July 1995.
29 This was a private member's bill which had a somewhat lukewarm support from the Government. In a previous attempt, the bill had been rejected, but a new energy efficiency minister (Robert Jones) managed to convince many of his colleagues of its benefits.
30 *ENDS Report*, No 254, March 1996, p 22.

they can obtain the data as auditing of private property has been specifically excluded from the Act.[31] Nevertheless, the plans may become important levers for future energy efficiency programmes.

As already mentioned, UK local authorities have limited scope for influencing public transport provision. Before bus deregulation, most local authorities owned bus services and could to some extent plan public transport provision. Since deregulation, the only way they can influence the services run by private companies is through dialogue with the companies and by providing subsidies, which in times of budget constraints can be difficult. This is not a very satisfactory situation and the kind of integrated transport planning found in Germany or Switzerland is almost impossible to achieve in the UK. There is thus an urgent need for central government to provide local authorities with greater transport powers, such as the possibility to set standards for local bus services.

CONCLUSIONS

While at the time of writing it appears likely that the UK will achieve its CO_2 emission stabilization target for 2000 (and possibly even some further reductions), there is nevertheless no coherent and comprehensive strategy to reduce emissions and also budget cuts have affected some of the measures listed in the CCP. Emission reductions are almost entirely attributable to fuel switching in electricity generation. Meanwhile, in the areas of transport, CHP and end use energy efficiency, little is being achieved. Renewable energy sources are being developed but at a slow pace. Hence, post 2000 emission increases are almost certain under a business-as-usual scenario.

A main problem in developing a coherent policy has been the Government's preoccupation with privatization and the introduction of competition in both the energy and transport sectors. In the process, in particular as a result of the establishment of quasi-autonomous regulatory authorities, the Government has lost much control over the industries. While there have been some incidental benefits associated with privatization and liberalization, some of the developments have also been negative in environmental terms. The regulators have not been given a strong environmental remit and certain issues, such as energy efficiency, clearly fall between regulatory chairs.

The energy efficiency issue exemplifies how environmental concerns continue to be ignored in government decision making, despite commitments in the Environment White Paper to ensure a better integration of environmental and other policies. A main problem is that, as in other Member States, the Department of the Environment has a relatively weak position within the ministerial hierarchy, especially in comparison to the Department of Trade and Industry which is responsible for most energy

---

31 Interview, Association for the Conservation of Energy, March 1995.

matters. Environmental measures have the best chance if they happen to fit in with the Treasury's priorities (ie budget cuts), as in the case of VAT on domestic fuels and cuts in the road programme. In fact, these measures were budgetary measures in an environmental disguise. Incidental benefits are not necessarily a bad thing but they are not sufficient and there is little evidence that environmental concerns ever receive priority treatment.

Analysis of the UK situation also indicates plenty of opportunities for emission reductions if the Government was prepared to intervene in appropriate ways. Clearly, energy efficiency, CHP and public transport are all areas where public policies could yield emission reductions. Several opportunities have already been missed, such as the failure to include energy efficiency provisions in the gas bill or to replace the FFL by a levy for energy efficiency purposes. John Gummer, the Secretary of State for the Environment, announced before the Berlin Conference of the Parties that the UK is prepared to consider emission reductions of 5 to 10 per cent post 2000. However, with the current policy inconsistencies, it is not clear how these can be achieved. One important factor may be that the current Conservative government may not be re-elected at the next general election, which has to take place by May 1997.

The opposition Labour Party is currently undergoing a major reform process under the new leader, Tony Blair. While a new Labour government would certainly instigate some policy changes, it is not clear whether it would be more pro-active on environmental measures. Environmental and sustainability issues hardly feature in Labour's economic policy discussions so it is not clear whether a Labour government would implement a strengthened climate change policy. Nevertheless, a relatively ambitious renewables target has been set (10 per cent of electricity demand by 2010, 20 per cent by 2025) and energy efficiency is to be taken much more seriously (Labour Party, 1994).

Overall, much remains to be done in the UK to implement a coherent and comprehensive climate change strategy with a long term outlook. Priority areas that need to be tackled are energy efficiency and transport, but the current government prefers to rely on market forces, even though to date these have failed to work. Little change can be expected until the next election, which is likely to take place in 1997. However, even then it is unlikely that climate change will be a priority area for a new government. Meanwhile, under a business-as-usual scenario, CO_2 emissions are set to rise after 2000.

REFERENCES

Agyeman, J and Evans, B (1994) *Local Environmental Strategies*, Longman, Harlow

Boardman, B (1993) 'Energy efficiency, the first or fifth fuel', Talk given at University of Hertfordshire, 6th December 1993

Cogen Europe (1995) *The Barriers to Combined Heat and Power in Europe*, Cogen Europe, Brussels

Collier, U (1994) *Energy and Environment in the European Union*, Avebury, Aldershot

Collier, U (1995) *Privatization and Environmental Policy: the UK Electricity Sector in a Changing Climate*, Working Paper 95/2, Robert Schuman Centre, European University Institute, Florence

Department of Energy (1990) *Energy Paper 58: An Evaluation of Energy Related Greenhouse Gas Emissions and Measures to Ameliorate Them*, London, HMSO

Department of the Environment (1995) *Climate Change: Update on the UK's CO$_2$ Programme*, issue 4, spring 1995, DoE, London

Department of the Environment (1996) *Climate Change: Progress Report*, DoE, London

Department of Trade and Industry (1994) *Energy Paper 62: New and Renewable Energy: Future Prospects in the UK*, HMSO, London

Department of Trade and Industry (1995) *Energy Paper 65: Energy Projections for the UK*, HMSO, London

Department of Transport (1989) *National Road Traffic Forecasts (Great Britain)*, HMSO, London

Department of Transport (1996) *Transport – The Way Forward: The Government's Response to the Transport Debate*, HMSO, London

Elliott, D (1994) 'UK renewable energy strategy', *Energy Policy*, Vol 22, no 12, pp1067–1074

EST (1994) *Strategic Plan 1993–2000, First Year Review, Corporate Business Plan 1994–1996*, EST, London

ETSU (1996) Renewable obligation status summary, obtained through personal communication

Friends of the Earth (1994) *The Climate Resolution*, Friends of the Earth, London

HM Government (1990) *This Common Inheritance*, HMSO, London

HM Government (1994) *Climate Change: The UK Programme*, HMSO, London

House of Commons Energy Committee (1992a) *Second Report, Consequences of Electricity Privatization*, HMSO, London

House of Commons Energy Committee (1992b) *Fourth Report, Renewable Energy*, HMSO, London

House of Commons Transport Committee (1995) *Fourth Report, Railway Finances*, House of Commons paper 1994–1995 206-I, HMSO, London

Labour Party (1994) *In Trust for Tomorrow*, Labour Party, London

Leicester City Council (1995) *The Leicester Energy Strategy*, City Council, Leicester

Newcastle City Council (1992) *Energy and the Urban Environment Strategy for a Major Urban Centre, Newcastle upon Tyne, UK*, City Council, Newcastle

O'Riordan, T and Rowbotham, E (1996) 'Struggling for credibility: The United Kingdom's response', in O'Riordan, T and Jäger, J (eds) *Politics of Climate Change: A European Perspective*, Routledge, London

Owens, S (1995) 'From predict and provide to predict and prevent?: Pricing and planning in transport policy', *Transport Policy*, Vol 2, no 1, pp43–49.

Parry, M and Duncan, R (eds) (1995) *The Economic Implications of Climate Change in Britain*, Earthscan, London

Rose, C (1990) *The Dirty Man of Europe*, Simon and Schuster, London

RCEP (1995) *Eighteenth Report: Transport and the Environment*, HMSO, London

Round Table on Sustainable Development (1996) *First Annual Report*, Department of the Environment, London

South Glamorgan County Council (1995) *South Glamorgan Transport Strategy, Final Consultation Draft*, South Glamorgan County Council, Cardiff

Chapter 7 | ITALY: IMPLEMENTATION GAPS AND BUDGET DEFICITS

Gianni Silvestrini and Ute Collier

INTRODUCTION

While Italy played an important role during its EU presidency in 1990 in helping to foster agreement on the CO_2 emission stabilization target for the year 2000, its domestic policy has not yet addressed the climate change issue in a comprehensive and coherent fashion. The National Energy Plan (NEP) (1988), prepared in 1988 in response to the nuclear referendum, when the climate change risk was not yet a topical issue, forecast a 13 per cent increase in CO_2 emissions by 2000 compared to 1987 values. Following Italy's signature of the FCCC, a National Programme (The First National Communication, FNC) for limiting carbon emissions was prepared in 1994 (FNC, 1995). However, this does not envisage emission stabilization – limiting the CO_2 emission increase to 2 per cent above 1990 values is considered the best that can be achieved in the year 2000. Italy has justified this position by referring to its energy efficiency record; it is the EU member state with the lowest energy intensity and one of the lowest CO_2 emissions per unit of GDP. The lack of indigenous fossil fuels and the rejection of nuclear power have provided a major economic incentive for energy savings. However, Italy is still the third largest CO_2 emitter in the EU and fast growth rates in demand for electricity and transport are likely to lead to increases in energy intensity.

Even if officially the stabilization target for the year 2000 is considered difficult to achieve, recent emission statistics show encouraging trends. In 1994, CO_2 emissions were 2 per cent lower than in 1990. However, this development is mainly due to the weakness of the Italian economy in recent years, rather than to an emission reduction strategy. In general, the measures presented in the climate change programme had already been planned or have been decided on the basis of other considerations, and a number have not been implemented yet, mainly because of budget constraints. This situation is likely to continue for the foreseeable future and a reduction of emissions by 2005 or 2010 seems quite unlikely.

GREENHOUSE GAS EMISSION TRENDS

As elsewhere in the EU, CO_2 is the most important greenhouse gas in Italy. Taking a 100 year time horizon, it accounts for 74 per cent of the radiative forcing, followed by CH_4 with 18 per cent and N_2O with 7 per cent, as is shown in Table 7.1.

Table 7.1. *Contribution of Different Emissions to Radiative Forcing in 1990*

Substance	Net emissions (Mt)	Contribution to radiative forcing at different time scales		
		20 years	100 years	500 years
CO_2	387	387	387	387
CH_4	3.9	242	96	29
N_2O	0.1	36	39	22
Total		665	522	438

Source: FNC, 1995

However, taking a 20 year time horizon, CH_4 is also very important, although the values in Table 7.1 have to be treated with caution, given the large uncertainty level still present in the estimates. The largest share of these emissions, 50 per cent, comes from agriculture, mainly from enteric fermentation and slurry. A further 41 per cent result from landfill sites and sewage, while leakages from natural gas extraction, transport and distribution account for the remaining 9 per cent. As far as nitrous oxide emissions are concerned, the largest contribution derives from the use of fertilizers (60 per cent) and from fossil fuel combustion (20 per cent).

Most CO_2 emissions derive from the energy sector, estimated at 401.4 Mt in 1990. To this total, a further 27.6 Mt of CO_2 derived from industrial processes (mainly from cement production) and 3.7 Mt from forest fires should be added. Carbon sinks are estimated at 40.4 Mt, which brings total net emissions to 391 Mt of CO_2. This figure is slightly higher than that shown in Table 7.1, due to a small revision in the latest update from the FNC.

Analysing the different sectors, the largest contribution is attributed to the transformation of energy (mainly electricity production), followed by the transport sector, as shown in Figure 7.1.

The contribution of the residential sector to overall emissions has declined steadily since 1984, with its share of total emissions much lower than in Northern European countries because of lower heating requirements. A fall in emissions can also be observed in the industrial sector, down from 28 per cent at the end of the 1970s to 23 per cent in 1993. Meanwhile,

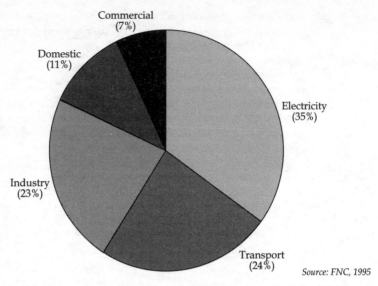

Figure 7.1: Sectoral contributions to CO_2 emissions in Italy (1990)

emissions from electricity generation and transportation have been growing continuously. As far as forest sinks are concerned, there has been a continuous increase over the past 30 years due to reforestation policies and to the natural regeneration of trees on abandoned pastures and arable fields.

Electricity generation (excluding imports) is now responsible for 35 per cent of the Italian energy-related CO_2 emissions (27 per cent of the total), amounting to 122 Mt in 1990. The largest part of the emissions, 87 per cent, are produced by the national utility, ENEL (*Ente Nazionale Energia Elettrica*), while the remainder are produced by municipal utilities and independent producers. As Table 7.2 shows, the Italian electricity sector is currently heavily dominated by oil.

Table 7.2. *Fuel Shares in Electricity Generation (1993)*

Fuel type	Share in electricity generation (%)
Oil	51.8
Hydro	18.9
Gas	18.0
Coal	9.4
Geothermal	1.7
Other renewables	0.2

Source: IEA, 1995

As discussed in more detail on pp115–116, this balance is now undergoing some major changes. However, at least in the short to medium term, this shift is likely to be towards gas rather than CO_2-free resources, although the share of renewables is expected to grow slowly.

CO_2 emissions have shown a small increase over the past 20 years, with the 1994 level only 2 per cent higher than that of 1978, as well as 2 per cent lower than that of 1990. This is due to a marked reduction in energy intensity and to changes in fuel shares, with an increase of natural gas use and a reduction of coal use. Emissions related to the use of natural gas have increased from 21 per cent to 24 per cent of total energy emissions between 1988 and 1994, while in the same period coal related CO_2 emissions have decreased from 15 per cent to 11 per cent.

Per capita CO_2 emissions in Italy (7.2 tonnes) are well below the EU average (9.2 tonnes). This is explained by a number of factors, such as the temperate climate, high energy taxes, limited indigenous energy resources and the high density of urban areas, the latter especially influencing transport emissions. In 1990, Italy had the lowest energy intensity (measured in toe/GDP) of all OECD countries except Switzerland (IEA, 1995). In terms of CO_2 emissions per unit of GDP in the EU only Austria, France and Sweden have lower ratios, due to their reliance on non-CO_2 energy sources (hydro and nuclear). Figure 7.2 shows Italy's position compared with the EU average.

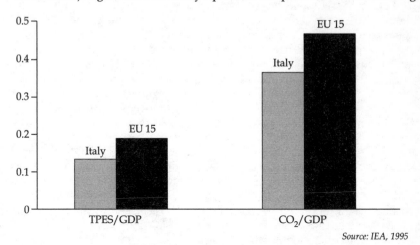

Source: IEA, 1995

Figure 7.2: Energy and CO_2 intensity in Italy compared with the EU average (1993)

However, energy intensity has seen little change since the mid 1980s, primarily due to lower fossil fuel prices which have reduced the scope for cost effective investment in energy efficiency (IEA, 1995). Furthermore, despite relatively favourable emission characteristics, Italy is nevertheless the EU's third largest emitter in terms of absolute volumes. Consequently, emission developments in this country are crucial to the overall emission picture of the EU. According to Italian Government projections, CO_2 emissions are expected to increase by around 3 per cent, although recent European Commission projections have adjusted this figure to 6 per cent (European Commission, 1996). There is thus little reason for complacency in Italy's response to the climate change issue.

THE ITALIAN CLIMATE CHANGE PROGRAMME

The Climate Issue and Italian Environmental Policy

Italy is generally considered one of the 'laggard' countries in environmental policy (Pridham and Konstadakopulos, 1994) and is notorious for having the worst record among EU Member States in implementing EU directives. However, environmental issues have become increasingly important in recent years, especially at local authority level. Indeed, Italy is the only country in the EU where a number of local councils (including Bologna, Florence, Milan and Rome) regularly impose traffic bans to deal with local air pollution problems. Furthermore, at national level, the Environmental Protection Agency has recently been established and a sustainable development (Agenda 21) programme was adopted in 1994. Italy thus appears to be becoming more proactive in environmental policy matters. However, as Marchetti (1996) has pointed out, while environmental laws and programmes have been adopted at an unprecedented rate, implementation is quite another matter. While the problem of implementation failure in the Italian political system has been well recognized, little has been done to make amends. As is shown later, in terms of activities relevant to climate change, the budget deficit and subsequent down scaling of many planned actions has been detrimental.

Climate change itself did not appear on the Italian political agenda until 1990. However, the 1987 nuclear referendum, in the wake of the Chernobyl accident, brought attention to the general effects which the energy system has on the environment. As a result, the nuclear programme was stopped and existing plants shut down. This had a somewhat negative impact on CO_2 emissions, although these have been limited as much of the generation shortfall has been compensated for by increased electricity imports, paradoxically from France and Switzerland, which themselves rely on nuclear power. More importantly, the rejection of nuclear power gave a new impetus to the promotion of energy efficiency, cogeneration and renewable energies, which are all beneficial in CO_2 terms.

The EU's initial climate change discussions coincided with the Italian Council Presidency in 1990. The then Environment Minister Giorgio Ruffolo played an important role in forging the decision by the Member States to collectively stabilize carbon emissions by 2000 at the 1990 level. During the 1991–1995 period, Italian international presence has been at a lower profile, but Italy usually supports the adoption of actions aimed at CO_2 reductions. Furthermore, Italy has been in favour of the EU carbon/energy tax, although for revenue rather than environmental reasons.

The early dynamism in ecodiplomacy has not been accompanied by a real national commitment to deal with this issue. Environmental groups, after an initial period of indecision (mainly due to the fear of a revival of nuclear power) strongly pushed the government to act. In 1991 *Legambiente*, the most active group in this field, launched a campaign to reduce carbon emission by 20 per cent by the year 2000 for which it collected 200,000 signatures. At national level, the climate change issue has been pushed by the Environment Ministry, usually in opposition to the Industry Ministry. Other Government ministries have hardly been touched by this issue. This is particularly true of the Transport Ministry, whose decisions continue to be taken without any consideration of the implications for greenhouse gas emissions. The Climate Change Programme itself was elaborated upon by a Ministry of the Environment working group, the so-called Commission on Climate, which included officials from other ministries.

Regarding other institutional levels, the Regions have reacted poorly to the climate change issue, even though the decentralization of the 1970s gave them responsibility for drawing up their own energy policies. However, the situation is more positive at the local level, where a certain involvement exists and a pilot programme on 'Cities against global warming' has been launched, as discussed on pp122–124. The next section examines the content of the national programme in more detail.

The Italian National Programme

In accordance with the requirements of the FCCC, ratified by the Italian Parliament on 15th January 1994, the Ministry of Environment drafted, in coordination with other Ministries, a report that was approved by the Interministerial Committee on Economic Planning (ICEP) on 10th January 1995. The report, following FCCC guidelines, contains:

■ An inventory of emissions and sinks of greenhouse gases for 1990.
■ The measures to be adopted in different sectors in order to limit CO_2 emissions to 1990 levels by the year 2000.
■ The vulnerability of Italian territory to climatic changes and adaptation measures.
■ Finance and technology.
■ Research and information.

As already pointed out, the measures listed in the programme do not actually lead to a stabilization of CO_2 emissions, but to a 2.9 per cent increase in emissions. To justify this lower objective, the programme refers to the already high efficiency of the Italian energy system, which makes further improvements very costly. At the same time, a clear message is sent in favour of the joint implementation of activities with Eastern Europe and developing countries, as a more effective way to achieve global carbon emission reductions. However, the pessimistic assessment of the economic burden of a CO_2 reduction policy is surprising, considering that the FNC is based on official studies which highlight the scope for reducing CO_2 emissions at negative costs. Different studies were conducted by ENEA (National Agency for New Technologies, Energy and the Environment) in order to define the CO_2 reduction potential and the cost of the different actions (Tosato and Contaldi, 1994; Onufrio, 1993; Contaldi, 1993).

Before analysing in detail the policies and measures presented in the FNC, some general remarks need to be made. Firstly, most of the measures presented had already been planned or were decided on the basis of other considerations. Yet, during the past five years efforts in the fields of energy saving and of renewable energy, on which the report relies heavily, have been less ambitious than originally planned and the report seems somewhat unrealistic in its assumptions in these areas (see below and pp 118–120). A second comment is related to the possibility of actually implementing the objectives presented. In many sectors, at present, this seems quite unlikely due to the lack of economic and regulatory instruments. Finally, it is rather difficult to evaluate the likelihood of the different reduction scenarios because there is not a detailed quantitative presentation of the actions to be taken. Additionally, the projections do not go beyond the year 2000. However, based on current trends, a reduction of CO_2 emissions in the years 2005 and 2010 seems quite unlikely. According to recent projections by oil companies, energy consumption in Italy is expected to increase by 16 per cent in 2010 compared to 1995 (UP, 1996).

ENERGY MEASURES: REDUCING FOSSIL FUEL DEPENDENCY

Energy Efficiency Programmes

Energy efficiency has been a priority in Italian energy policy since the oil price shocks of the 1970s, on account of Italy's lack of indigenous fossil fuel reserves. Although environmental issues have not played a central role in Italian energy policy, this emphasis on energy efficiency has had beneficial effects, contributing to the low carbon intensity of the Italian economy. The NEP, elaborated in 1988 in the aftermath of the 1987 nuclear referendum, further emphasized the role of energy saving and, for the first time, included specific targets for the renewable sources. In terms of CO_2 emissions, the NEP projected an increase of 13 per cent in 2000 compared to 1987. However, confirming previous experiences in Italy, central planning has turned out to be quite ineffective, and the situation during the past seven years has diverged considerably from the plan. In particular, the share of coal has not grown as planned, energy efficiency actions have been limited due to budget cuts and the share of 'new' renewable energy remains marginal.

In 1991, as part of the NEP, a specific law (n. 10) was issued to promote energy efficiency both for generation and for end use. The high expectations raised by this law were subsequently shattered, as the formal acts accompanying the law have suffered long delays (some of them have not yet been approved) and as the initial budget (2500 billion lire, nearly 2.2 million ECU) has continuously been reduced, as Figure 7.3 demonstrates. The budgets for

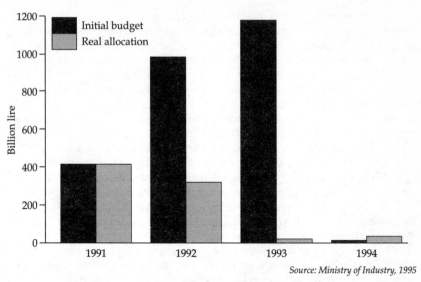

Source: Ministry of Industry, 1995

Figure 7.3: Funds allocation for energy efficiency (law n. 10/1991)

1993 and 1994 were very modest, making Italy's the smallest budget for government energy efficiency programmes in the whole of the OECD.

The allocations for 1995 and 1996 were 81 and 304 billion lire, with the latter figure suggesting that this policy measure might be awarded higher levels of funding by the new government.

Not only have there been budget problems, but the allocation of funds between different sectors and the decision making mechanism have also been criticized. Half of the incentives were directed to cogeneration plants and district heating systems, limiting other promising options. It is, however, interesting to note that even though the incentive policy has been very limited, the efficiency market has grown in recent years, in spite of difficult economic conditions. A study – that examined 450 companies which purchased eight different types of energy saving products – showed sales of 4.3 billion lire (1992 value) with an average yearly growth rate in the 1988–1992 period of 14 per cent (Agostini and Sirtari, 1995). Only 7 per cent of the products sold used some form of financial incentive. According to the industries involved, this form of government help is irrelevant and other forms of intervention should be proposed.

High Energy Taxes and Electricity Tariffs

As a result of the problems with the energy efficiency incentive programme, further improvements will have to rely mainly on market forces. Here, price levels are important. Indeed, one of the reasons for the lower energy intensity of the Italian system compared with other European countries is the high level of energy taxation applied in this country. However, the beneficial effect on energy efficiency has been largely coincidental, as the main aim of taxation was to increase state revenues, to control inflation and to reduce budget deficits. According to Paoli (1996), tax revenue linked to energy products currently constitutes more than 15 per cent of state tax revenue. The promotion of energy conservation and the protection of the environment has not been a serious objective of Italian fiscal policy. This can be seen, for example, by the fact that coal, the most polluting source of energy, has a low level of taxation. The domestic sector has a much higher level of taxation, while industrial sector tariffs are more in line with international levels in order to preserve the competitiveness of Italian industry. The sector on which fiscal policy is focused most is the transport sector. The cost of petrol in March 1995 was 447 lire, while the selling price was 1855 lire. The high fuel prices are one of the explanations for the high level of efficiency of the Italian car fleet. However, as shown on pp 120–122, high fuel prices have not stopped the rapid expansion in car ownership and CO_2 emissions from the transport sector are growing fast.

Residential sector efficiency should be influenced by new electricity tariffs introduced in 1993, with drastically increased progressivity. For a monthly consumption larger than 220 kWh, the cost reaches 0.26 ECU/kWh, compared with an average price of 0.07 ECU/kWh. However, the national utility ENEL, in contrast with different municipal utilities, has not provided consumers with any information regarding the new tariff structure and generally does little to reduce the information deficit, a well-known obstacle to better energy efficiency.

As far as a carbon tax is concerned, the Government in 1994 proposed an increase of $7 per barrel of oil equivalent (aligned with the original EU carbon tax of $10) to be deducted from other taxes to ensure revenue neutrality. However, the proposal has not progressed, but the new government intends to revive the carbon tax and partially use it for environmental projects. Similar and more radical proposals have been elaborated by environmental groups. Generally, it is clear that, despite high prices, there is still much potential to improve energy efficiency in Italy, which suggests the need for other measures such as mandatory standards.

Changes in the Electricity Sector

Electricity is the sector, together with transportation, in which CO_2 emissions are still increasing and will continue to grow in the future according to the official scenarios. While policies to manage electricity demand have been limited, there are signs of changes in the production side that will lead to a reduction in CO_2 emissions. According to the NEP (1988), electricity demand in 2000 was expected to reach between 290 and 315 TWh. More up-to-date projections from ENEL (1995) suggest a consumption of 315 TWh in 2002, with an annual growth rate of 3.5 per cent, while according to oil companies the consumption will be 337 TWh in 2005 and of 370 TWh in 2010 (UP, 1996). According to the ENEL study, capacity requirements will grow to 73 GW by 2010 (an increase of 50 per cent from 1994 capacity), with an increase in the share of coal and gas, as well as a possible nuclear revival of up to 4 GW. Notwithstanding the highly unrealistic assumptions regarding nuclear power, CO_2 emissions would almost double from 1990 (120 Mt) to 2010 (215 Mt). The conclusions of the report state that:

> *such projections are strongly diverging from the theoretical stabilization target at the 1990 level and show that in Italy this is unrealistic (ENEL, 1995).*

However, based on current trends and the changes taking place in the electricity sector, the situation regarding emissions looks somewhat more positive. Strong local opposition has almost totally stopped the construction of new coal fired power plants. Conversely, there is a rapid growth of independent producers who have taken advantage of the provisions under law 9 of 1991, which liberalizes electricity generation for CHP and renewables plants and sets a preferential price for the sale of surplus electricity for the first eight years of generation. ENEL has been allowed to pass the extra costs on to its consumers. Also, government grants for 40 per cent of the capital costs of CHP plants were supposed to be offered, but these were again hit by budget problems. Nevertheless, after a period of continuous decline in production by independent producers post 1979, there has been a complete trend reversal since the late 1980s. Given the favourable new conditions, applications for the licensing of 10,000 MW (mainly gas fired cogeneration plants) have been presented, of which 6700 MW have been authorized by the government. Independent production should grow from 17 per cent of the total in 1994 to 30 per cent in the year 2000 (from 37 to 90 TWh).

ENEL itself is also making some major changes to its generation facilities and has adopted, for the period 1994–2000, a series of modifications to the original plans presented by the NEP (1988), with investments in the order of five million ECU:

■ Closure of 3500 MW old power plants with an average efficiency of under 34 per cent.
■ Conversion to a combined cycle of 1600 MW of existing plants and construction of 1800 MW new combined cycle plants with efficiencies of 45–50 per cent.
■ Construction of 3100 MW of new conventional power plants with efficiencies of 40 per cent.

According to the latest ENEL and independent producer plans, the efficiency of fossil fuel electricity generation should rise from 37.8 per cent in 1990 to 39 per cent in 2000 and then to 40 per cent in the following years. In terms of specific carbon emissions, these decisions will lead to reductions from the value of 550 g/kWh of 1990 to 513 g/kWh by the year 2000 (–7 per cent). Overall, the changes in the power production system and in the fuel mix should lead to a 3 Mtoe per annum reduction in energy demand and to a 9 Mt per annum reduction in CO_2 emissions (Carta et al, 1992a).

Furthermore, significant improvements in end-use efficiency (and hence

CO_2 reductions) could be achieved with a more active role of the electric utilities and with the introduction of innovative saving strategies. Considering the difficulties involved in building new capacity due to local opposition to coal power plants, the Italian situation is particularly suitable for DSM programmes. Nevertheless, there have been few results in this field. Between 1983 and 1992 a number of programmes focusing on solar water heaters, power factor correction, heat pumps and compact fluorescent lamps were in operation at a cost of 122 million ECU (less than 1 per cent of ENEL's annual revenue), resulting in savings of 364 ktoe per annum. ENEL estimates the possible savings by the year 2000 to lie between 5 TWh per annum and 21 TWh per annum (Carta et al, 1992b) but has currently no plans for further programmes.

Overall, the prospects for emission reductions in the electricity sector are mixed. On the one hand, growth in electricity demand has been lower than expected so the entry of new independent generators is leading to overcapacity in the supply system. On the other hand, liberalization of the electricity market with the planned privatization of ENEL could reduce the potential for DSM. The current situation regarding the exact nature of privatization and liberalization is uncertain. The newly established regulatory authority for the energy sector has been given the responsibility to ensure the integration of environmental concerns, but it remains to be seen how seriously this will be taken. Recent decisions, like the appointment of an environmentalist to the presidency of ENEL and the possibility of recovering investments in DSM through the proposed 'price cap' tariff mechanism, give some cause for optimism.

The activities of municipal utilities are supposed to grow within the framework of the liberalization process. An interesting example is given by a consortium of municipal utilities that is being promoted by ENEA in order to enlarge the activities in the field of DSM and of renewable energy. Local municipal utilities have generally become more active in recent years, as the example of Acea, which operates in Rome, shows. A compact fluorescent lamp programme, promoted in 1995 in cooperation with Greenpeace, resulted in additional sales of 100,000 units, with an estimated saving of 2.7 GWh per annum and a 3 MW peak reduction (Acea, 1995).

Renewable Energy: Slow Progress

Renewable energies already make a significant contribution to energy production in Italy, amounting to around 7 per cent of primary energy requirements, mainly based on hydropower and geothermal power, for which Italy has the largest resource in Europe. Nevertheless, there is much scope for the development of other renewable energies, which in view of Italy's lack of indigenous fossil fuel reserves should be a top priority. The NEP (1988) for the first time set clear targets for energy production from renewable energies to be achieved by the year 2000:

■ 300–600 MW of wind energy (1–2 TWh);

■ an additional 1.5 Mtoe from biomass;
■ an additional 10 TWh from hydroelectricity;
■ an additional 6 TWh from geothermal power.

Overall, electricity generation from renewable energy by 2000 was supposed to increase from 48 TWh to 69 TWh, amounting to a 44 per cent increase. Furthermore, for solar energy 2×10^6 m^2 of hot water collectors were planned in addition to the 400,000 in place in 1987, while for PV plants an intermediate target of 25 MW was considered achievable for the year 1995. In order to achieve these goals, two forms of incentives were provided – a financial grant to cover part of the capital cost of the technologies and a special rate for the sale of electricity to reflect the environmental benefits. However, it was only at the end of 1992 that the rules for the sale of electricity from renewables and from cogeneration were established.

The investments considered in the NEP (1988) in order to develop these technologies were in the order of 4.2 billion ECU for solar, wind and biomass, 1.7 billion ECU for geothermal and 7.5 billion ECU for hydroelectricity. As far as the incentives included in law n. 10/91 (1.2 billion ECU) are concerned, there has been a considerable shortfall due to budget constraints, so there has not been sufficient funding to finance the law. Also the special rates for renewables have been considered too low by potential investors, in particular for wind energy and for PVs.

These points help to explain the slow pace that has, until now, characterized the diffusion of renewable energy in Italy. At the end of 1994 there were only 22 MW of wind power and 14 MW of PV systems in operation. As far as geothermal energy is concerned, Italy is quite advanced with an installed electrical capacity of 626 MWe and 852 MWe projected for 2002, while thermal capacity amounts to 700 MWth. However, the situation could rapidly change, at least in three sectors, namely wind power, waste to energy and solar thermal. Several wind farms are planned, and bids for 1000 MW of wind turbines were presented to the Ministry of Industry in late 1995. This increased interest is a result of some important changes in the framework conditions:

■ An increase in the special rate for wind energy generated electricity (184.5 lire/kWh for the first eight years).
■ A better evaluation of the wind potential, in particular for internal areas.
■ The possibility of using the 1994–1999 structural funds of the European Union.
■ A more active role of ENEL, which until 1995 was very cautious regarding this technology.

The two existing Italian wind energy companies are trying to make up for their late entry into the market, but the arrival of foreign capital and technologies seems to be the real factor that has changed the wind perspective in Italy. Given Italy's high population density, the main obstacle to the diffusion of wind farms will be their visual impact.

In waste to energy production, high disposal costs, as well as the increasing difficulty in disposing of urban wastes into landfill sites, favour the installation of new incinerators. Landfill sites have also been considered by the government for the production of energy. In this case, there is a combination of benefits from the point of view of the climate change issue. Since the emissions of CH_4 from landfills are relatively high in Italy, the capture of biogas and its use as fuel in power generation will greatly contribute to the reduction of greenhouse gas emissions. The target contained in the FNC is to burn, by the year 2000, 0.3 Mt of CH_4, with 100 MW of power plants installed.

Considering Italy's high solar radiation, the low rate of solar water heating is particularly deplorable. The current rate of 15,000 m^2 installed per year is a very modest achievement. However, there are some signs that the number of solar thermal panels could greatly expand after this decade of very low profile. An installation rate of 100,000 m^2 per year is considered a reasonable target for the near future. The past year has seen the combined efforts of a number of players in order to achieve this goal. These include ENEA, ISES (International Solar Energy Society), Eurosolar, single municipal utilities and local authorities. At the time of writing, ENEA was working on a manual on solar collector technology involving large scale diffusion. Courses for installers are being run to eliminate bottlenecks caused by the lack of professionals in this field.

TRANSPORT SECTOR DEVELOPMENTS

The transport sector accounted for 24 per cent of energy-related CO_2 emissions in 1990 and its share is growing. Road based car and lorry traffic accounts for nearly 90 per cent of these emissions. While, according to Danielis (1995), Italy has the lowest aggregate transport energy intensity in the OECD, the share of the transport sector in energy end use is nevertheless increasing (from 22 per cent in 1975 to 33 per cent in 1994). Hence, while fuel use per passenger km has fallen in all modes (except railways), with Italy having the most efficient car fleet in the OECD, this has been more than compensated for by growing traffic volumes, both in terms of the number of vehicles in circulation and distances travelled. Car ownership increased from 313 to 520 vehicles per 1000 inhabitants between 1980 and 1994 (Danielis, 1995). Italy now has the highest rate of car ownership in the EU. Furthermore, there has been a clear and growing imbalance in the modal split towards road transport (in particular in urban areas where the car passenger km share has increased from 85 per cent in 1985 to 93 per cent in 1993). While the projections made in 1985 by the National Transport Plan (PGT, 1988) assumed increases of 29 per cent for freight transport volumes and 26 per cent for passenger transport volumes between 1985 and 2000, the growth rates are now projected to be even higher at 45 per cent for freight transport and slightly less for passenger transport.

Transport is undoubtedly the most problematic sector in which to achieve CO_2 emission reductions. While during the 1970s and 1980s energy consumption and economic activity were decoupled in the industrialized countries, there are no real signs that a similar 'revolution' could be taking place in the transportation sector. On the contrary there are strong forces, such as the growth in freight transport under the Single European Market, which push in the opposite direction. However, if carbon reduction targets are to be achieved during the next decades, it is quite obvious that a similar decoupling must take place, breaking the link between mobility and economic growth and reducing the transport sector's growth rate.

Currently, in Italy, there is no clear strategic emphasis on the investments necessary to balance the transport modal split. Official documents, like the Italian Agenda 21 strategy and the FNC, consider this as an important target to achieve, but in practice actions are not as progressive as they could be. The targets for the year 2000 (with respect to 1991 levels) included in these documents are as follows:

■ Increase the share of urban transport expenditure from the actual level of 2–2.5 per cent of total transport infrastructure expenditure to the average of other European countries (6–9 per cent).
■ Increase the share of public transport in urban areas from 11 per cent to 20 per cent.
■ Increase the total length of subway systems by 25 km.
■ Construction of 1250 km of light rail (tramway) systems (25 km for each of 50 cities).
■ Increase passenger rail usage from 20 to 34 billion passenger km.
■ Increase rail freight transport from the actual 23 billion tonne km (12 per cent) to 42 billion tonne km.
■ Construction of 2000 km of cycle paths (20 km for each of 100 cities).

In reality these targets are unlikely to be achieved by the year 2000, primarily because much of the Government's transport budget has been allocated to a small number of large projects (the so-called *grandi opere*), without a strategic assessment of their environmental costs and benefits. An example is the debate over the *Variante di valico*, a second motorway between Bologna and Florence. This project is supposed to reduce traffic on the existing motorway, which is congested with freight transport. An alternative to this proposal would be an expansion in rail transport, which would also be consistent with the policies adopted by the border countries of Austria and Switzerland which have decided to introduce limits to lorry transit through their countries. By 2004, no lorries will be allowed to pass through Switzerland, as a result of a 1994 referendum, which obviously has implications for Italy. Meanwhile, a large part of rail investment in Italy has been allocated to high speed train lines (such as the extension of the line from Rome to Naples and a new tunnel crossing in Florence), while other services are being neglected. Another proposed project is a bridge over the Messina Strait to reduce the peripherality of Sicily. The transportation benefits would be minimal, while there are various environmental costs. Generally, it has been remarked that the powerful national car lobby (represented in particular by FIAT) has had some influence on the bias towards road based investments (Klimabündnis, Ökoinstitut Südtirol and Österreichisches Ökologieinstitut, 1995).

A further problem is institutional, in that the effective planning and financing of a transport infrastructure is hindered by the wide diffusion of responsibilities. There are 21 public powers involved in transport decision making, with much individual autonomy. The national Ministry controls only a third of the expenses in this sector. There is a lack of coordination between different levels, as well as a lack of consideration of climate change concerns. Nevertheless, on the positive side, after a long period of neglecting urban public transport, there are signs of a revival. Law 211 of 1992 provided 2 billion ECU to cofinance at 50 per cent the construction of rail based public transport in cities. In November 1995, after a long evaluation period, 15 projects in 11 towns were approved. Three-quarters of the investments will focus on the completion of subway systems, with two new light rail underground systems for the cities of Turin and Brescia. However, the light railway systems (39 km of tramway networks and 11 km of automated light rail) financed by this law represent less than 5 per cent of the target of 1250 km included in the FNC. New projects will be financed in 1996, but considering that the construction of new light rail systems takes three to four years, it is quite likely that in the year 2000 there will be no more than 20 per cent of the total originally planned. An analysis made for two projects for the city of Palermo has shown that the reduction of CO_2 per million ECU invested would be 60 per cent larger for a tramway option than for a subway. There thus appears a misallocation of resources due to a lack of consideration of environmental factors.

LOCAL CLIMATE CHANGE POLICIES

Creating the Right Organizational Framework

Local authorities are particularly important in the strategy to deal with transport emissions. All Italian cities with a population larger than 50,000 have to prepare an Urban Traffic Plan, with the reduction of energy consumption as one of its objectives. Many large towns are trying to discourage the use of private cars through the introduction of parking fees, pedestrian areas and periodic bans of all traffic. The main environmental concern in these policies is the reduction of air pollution but CO_2 benefits will automatically ensue. Furthermore, in 1991 a law was adopted compelling all regions, as well as towns with more than 50,000 inhabitants, to produce and adopt local energy plans (regional and municipal, respectively), with a special emphasis on renewable energy sources and cogeneration. Until now, however, only a few smaller cities (Padova, Rovigo) and some large towns (Turin, Rome) have fulfilled the law's requirements, since political will to enforce this law has not been shown by the Rome Government. Cispel (*Confederazione Italiana Servizi Pubblici Enti Locali*), the organization of municipal utilities, is trying to push for the introduction of energy plans at local level.

Hence, while in principle there is a framework that gives local authorities responsibilities in fields important for climate change abatement, once more implementation is a problem. However, there are signs that many towns and cities are becoming more active in this area. Many Italian cities are participants of ICLEI's Cities for Climate Protection Campaign, with nine members, including all the major cities (Bologna, Florence, Livorno, Milan, Naples, Palermo, Rome, Turin and Venice). A further 20 towns, mainly in the predominantly German speaking region of Alto Adige, are taking part in the Climate Alliance (see Chapter 4).

A first attempt to create a national network of towns involved in the definition of a strategy linking the solution of local environmental problems to more general risks, like global warming, was made in 1992. The stimulus came from Bologna's participation in ICLEI's Urban CO_2 Project and the coordination of the network was undertaken by the environmental group *Legambiente*. More than 20 cities formally agreed to participate, some workshops were held in Rome and a national congress was organized in Bologna. However, the programme faltered for two main reasons. Firstly, the instability of local governments did not allow any form of continuity of the planned activities. Secondly, the absence of a real coordination structure due to a lack of funds greatly reduced the impact of the project.

The situation has changed in the past few years. Thanks to a new electoral system the Italian city governments are now more stable. The 1993 municipal elections saw in many towns the victory of progressive forces with a clear environmental approach. Given the more promising framework, in 1995 it was decided to reconsider the project 'Cities against global warming', but with a different approach. Fewer towns were selected, focusing on those in which innovative energy plans were already in progress, or at least planned. The towns chosen are (from South to North): Palermo, Rome, Livorno, Bologna, Turin, Padova and Rovigo, all of different size, geographical position and structural characteristics. The focus in each of the cities participating in the project will be on projects with which they are already involved and on disseminating the results to the other cities participating in 'Cities against global warming'.

Some Encouraging Examples – Bologna and Palermo

In 1991 Bologna, located in central Italy, was selected along with 12 other towns in industrialized countries to participate in the Urban CO_2 Project organized to elaborate effective strategies to reduce carbon emissions. In order to define a carbon reduction target for the year 2005, energy saving potentials in different sectors were evaluated. A 'climate stabilization' scenario was defined, which considered various policy measures that could be implemented between 1995 and 2005 at the local, national and international level (Silvestrini, 1996). The results suggested that if a coherent reduction policy was implemented at all levels, a 31 per cent reduction of greenhouse gases would be feasible by 2005 (compared to 1990 levels). Based on the results of this study, the city authorities adopted a 30 per cent

CO_2 reduction target for the year 2005 in February 1995 (compared to the 1990 value). Specific local actions were to achieve an 8 per cent emission reduction. As a first measure, a 3 MW wind farm is being built in the mountains near the town and operated through the local municipal utility.

Meanwhile, the city of Palermo is developing an interesting solar power programme. One of the main obstacles to the diffusion of solar thermal systems is the inadequate training of installers, as well as the high prices of solar collectors in Italy. The strategy of the municipality of Palermo tries to deal with both these aspects. Given the good results of self-construction seminars in Switzerland and Austria, the idea was to proceed in this direction, with the aims of increasing local added value (and employment) and producing cheap solar components. Two one-year courses were organized by a technical secondary school and a workshop was organized by the Unions for 40 redundant workers. Some successes have already been noted. A group of technicians from the technical Institute has decided to create a solar company and already has secured some orders from the private sector. Furthermore, the municipality has financed a programme to install solar systems in 29 public buildings, involving employees who participated in the workshops.

CONCLUSIONS

Italy, together with Spain and Greece, is generally considered a laggard country in environmental policy, especially as far as implementation and enforcement are concerned. This chapter shows that the climate change issue is not a high priority in the Italian government agenda, at least internally. While Italy has been fairly proactive at the EU level and in the UN FCCC negotiations, domestically there has been little change to current policies and a reliance on past performance, which resulted in relatively low per capita and per unit GDP emissions. While an official climate change strategy has been published, its content is rather thin and based on unrealistic assumptions concerning the implementation of existing measures (IEA, 1995). Decisions taken by government ministries other than the environment ministry rarely consider this parameter in the choices that are made. The absence of a clear policy is particularly noticeable in the sectors with the largest and increasing share of carbon emissions; transportation and electricity.

Nevertheless, in both sectors some decisions have been taken which are favourable in emission terms, such as investments in efficient electricity generating plants and in public transport systems. While climate change considerations have played no role in these decisions, they are nevertheless beneficial. As far as energy efficiency and renewable energies are concerned, Italy has some laws with good provisions but the problem has been one of implementation, a general feature in Italian policy making. The laws were adopted at a time when budget constraints became ever more severe and these two areas have suffered accordingly. Also, government upheavals in recent years have hindered a consistent policy, with several environment ministers, not all of whom have taken this portfolio very seriously. A further problem has been the diffusion of responsibilities between central, regional and local governments, with a lack of coordination and a number of disagreements.

The Italian example is particularly interesting in so far as energy prices are concerned. High taxation has ensured that fuel prices in the domestic sector and in transportation are among the highest in the EU. While this has contributed to the relatively low energy intensity of the Italian economy, there are nevertheless plenty of examples of inefficiencies. Houses in North and Central Italy, where seasonal temperature differences make heating necessary for only part of the year, are poorly insulated and appliance efficiencies are well below the best available technology levels. In the transport sector, Italy has a relatively fuel efficient fleet but car ownership is now among the highest in Europe. These examples show that high prices on their own are not sufficient to cut CO_2 emissions, a fact which is an important indicator of the likely effectiveness of carbon taxes.

CO_2 emissions have actually decreased slightly since 1990, mainly because of a slow down in economic growth. Changes on the electricity generation side will also result in emission reductions. Overall, emission stabilization by 2000 might be achieved, although officially the government projects an increase of nearly 3 per cent. Generally, Italy's CO_2 balance is more favourable than that in most other EU countries due to the low levels of energy intensity. However, total emissions are high and a comprehensive policy is needed to achieve reductions in CO_2 emissions during the next 10–20 years. Given the current economic difficulties Italy is facing, it will continue to be difficult to finance large programmes that provide incentives for energy efficiency and the use of renewable energy sources. There is thus a need for a policy which focuses on market transformation tools, such as labelling, energy efficiency standards and DSM, to complement the energy taxes and tax shift policies. Given the resistance at national level to the formulation of such measures, the role of the EU is fundamentally important as an institution to give policy direction through binding instruments such as directives, in defining regulations, standards and in proposing carbon/energy taxes. Italy has generally been supportive of Commission proposals in these areas.

An important role in the definition of a climate change policy could be played by local authorities in combining the solving of local environmental problems with a contribution to the reduction of the greenhouse gas emissions. Here, some local authorities are already very proactive but there is a need for these activities to become more widespread. Some useful laws are already in place, but once again implementation and enforcement are lacking.

REFERENCES

Acea (1995) *Piano energetico ambientale del Comune di Roma*, Istituto Ambiente Italia, Roma

Agostini, M and Sirtori, L (1995) 'Il mercato dei prodotti di risparmio energetico', *Risparmio Energetico*, No 44

Carta G et al (1992a) *The Changing Italian Electricity Supply Industry : Impacts on the Future Generation System*, Proceedings from Cigré Session 1992, Paris, unpublished

Carta, G et al (1992b) *The Italian Generation System Evolution Up to 2000 and beyond – Problems, Actions, Expected Results*, Proceedings from Cigré Session 1992, Paris, unpublished

Cogen (1995) *The Barriers to CHP in Europe*, Cogen Europe, Brussels

Contaldi, M (1993) *Opzioni Tecnologiche per Ridurre le Emissioni di Anidride Carbonica nei Settori Industriali ad Alta Intensità Energetica e nella Produzione di Energia*, Internal report, ENEA, Roma

Danielis, R (1995) 'Energy use for transport in Italy', *Energy Policy*, Vol 23, no 9, pp799–807

ENEL (1995) *L'Attività dell' ENEL – 1994*, ENEL, Roma

European Commission (1996) 'Second evaluation of national programmes under the monitoring mechanisms of Community CO_2 and other greenhouse gas emissions', *COM* (96) 91

FNC (1995) *First Italian National Communication to the Framework Convention on Climate Change*, Ministry of Environment, Rome

IEA (1995) 'Italy', in *Energy Policies in IEA Countries*, IEA/OECD, Paris, pp305–345

Klimabündnis, Ökoinstitut Südtirol and Österreichisches Ökologieinstitut (1995) *Optimizing the Climate Protection Strategies of Local Authorities in Europe – Final Report*, Frankfurt/Main, Bozen, Wien, unpublished

Marchetti, A (1996) 'Climate change politics in Italy', in O'Riordan, T and Jäger, J (eds) *Politics of Climate Change: a European Perspective*, Routledge, London, pp298–329

Ministry of Industry (1995) personal communication

NEP (1988) 'Piano Energetico Nazionale', *Notiziario dell' ENEA*, No 8–9

Onufrio, G (1993) *Emissioni di Anidride Carbonica e Opzioni Tecnologiche per la loro Riduzione in Italia*, Internal report, ENEA, Roma

Paoli, L (1996) 'Italian energy policy: from planning to an (imperfect) market', in McGowan, F (ed) *European Energy Policies in a Changing Environment*, Physica Verlag, Heidelberg, pp 88–129

PGT (1988) *Il Trasporto Merci e L'economia Italiana*, Piano Generale dei Trasporti, Roma

Pridham, G and Konstadakopulos, D (1994) 'Sustainable development in Mediterranean Europe? The dynamics of environmental policy-making and interactions between European, national and sub-national Levels,' paper for the conference *The Politics of Sustainable Development*, University of Crete, 21–23 October

Silvestrini, G (1996) 'Il piano di riduzione delle emissioni di gas climalteranti della città di Bologna', in Legambiente (1995) *Ambiente Italia 1995*, Edizioni Ambiente, Milano, pp68–71

Tosato, G and Contaldi, M (1994) 'Opzioni tecnologiche ed amministrative contro i rischi dell'effetto serra', *Energia*, No 2

UP (1996) *Previsioni di Domanda di Energia e Prodotti Petroliferi in Italia*, Unione Petrolifera, Roma

*Pierre-Noël Giraud with Ute Collier
and Ragnar E Löfstedt*

Chapter 8 | **FRANCE: RELYING ON PAST REDUCTIONS AND NUCLEAR POWER**

INTRODUCTION

The following quote by an EU official serves as a good introduction to the attitude to the climate change problem in France: 'In Europe, as regards energy, there are two countries: France and the others'. The attitude of the French government towards the means and the ends of a policy aimed at preventing climate change is shaped by the particularity of the French energy system founded upon the energy policies implemented since the first oil shock.

France's reliance on nuclear power is unique in Europe, with this power source providing over 80 per cent of France's electricity generation and 40 per cent of her primary energy production. The switch from imported fossil fuels to nuclear power occurred in response to the first oil price shock and has led to an improvement in the level of energy independence from 22.5 per cent in 1973 to 51.6 per cent in 1994. Furthermore, together with other developments, it has also resulted in a significant, if fortuitous, reduction in CO_2 emissions from energy production of 25 per cent between 1973 and 1990. Of this reduction 62 per cent is due to the switch to nuclear power, 31 per cent to increased energy efficiency and a further 7 per cent is a result of industrial restructuring (IFEN, 1994). Notwithstanding the contentious nature of nuclear power, there is no denying that it has had a positive effect on France's CO_2 emission balance. CO_2 emissions in 1990 stood at 5.92 t per capita and 0.31 t per $1000 of GDP, making France one of the lowest CO_2 emitters in the EU (and the industrialized world as a whole) on a per capita and per unit of GDP basis. However, overall, France is still the fourth largest emitter after Germany, the UK and Italy and its emissions are far from negligible.

In accordance with its particular circumstances, the French government rejects flat rate emission reduction targets. This is based on the argument that the achievement of a percentage based reduction target, such as a 20 per cent reduction of 1990 emissions, would cost more in France than in other OECD countries, let alone in the Eastern European post Communist countries. Instead, the Government prefers policies which are defined in terms of a ceiling cost incurred by each tonne of reduced emissions. Consequently, it has proposed that all adherents to an eventual climate change protocol, and primarily those in Europe, should instigate policies to implement all measures that incur a cost of less than 70 ECU per tonne of CO_2 reduction.

Nevertheless, the French government has signed the FCCC and has drawn up a national programme containing a number of measures (Premier Ministre, 1995). As this chapter shows, these are generally a continuation of previous policies and not all are being implemented or they lack effectiveness. Overall, the French attitude to the climate change issue can be best characterized as indifferent; little effort is being made to reduce emissions from the transport sector, which, like elsewhere, are on a continuous upwards trend. Furthermore, little attention is being paid to more general issues of sustainability, especially in the energy sector, although there are signs that renewable energies will be taken more seriously in the future. With current emission trends, stabilization, let alone a reduction of CO_2 emissions, is unlikely. At the same time, it has to be acknowledged that France faces a difficult job in reducing emissions, with reductions needed primarily in the transport sector, notoriously difficult to tackle.

FRENCH GREENHOUSE GAS EMISSIONS: RELYING ON PAST REDUCTIONS

As Table 8.1 shows, by far the most important greenhouse gas in France is CO_2. Using global warming potentials (GWP), CO_2 (minus sinks) accounts for 74 per cent of total emissions, while CH_4 (as the second most important gas) contributes 14 per cent and N_2O 12 per cent. Given the projected increases in CO_2 emissions (22 Mt by 2000 according to the national programme, mainly in the transport sector) and the planned stabilization and reduction of most other gases, the relative importance of CO_2 is set to increase further.

Currently, the main sources of CH_4 are from agriculture (55 per cent), waste disposal (25 per cent), fuel consumption (15 per cent) and gas networks (10 per cent). Research into CH_4 reductions within the agricultural sectors has only recently begun, and the current view is that there will not be any major reductions in the foreseeable future. Within the waste disposal sector, plans have been drawn up to reduce CH_4 from landfills via a larger incineration programme, while measures to recover CH_4 from abandoned landfill sites are only in the research stage. Most of the planned CH_4 reductions will occur in the gas pipeline sector, where there are plans to

Table 8.1. *Total French Greenhouse Gas Emissions for 1990 and 1993*

Gas	1993 (in Mt)	Variation (1993/1996) (%)
CO_2 emissions: all sectors:	365	–0.5
Energy consumption	351	+0.3
Industrial processes	14	–15
CO_2 absorption by soil and forest	–37.2	–15.5
Total net CO_2 emissions	327.8	–2.1
CH_4 (equivalent CO_2)	2.83 (70.7)	
N_2O (equivalent CO_2)	0.171 (46.1)	
NO_x	1.675	
Volatile organic compounds	2.3	
CO	10	

Source : Premier Ministre, 1995

replace a total of 1600 km of old cast iron pipelines with modern steel and polyethylene pipes. This will lead to major reductions as the old style pipes are ten times as leaky as the modern ones. The two largest sources for N_2O are industrial processes (60 per cent) and agriculture, particularly artificial fertilizers (35 per cent). According to government sources, emissions in the industrial sector will decrease by 75 per cent by the year 2000 through stringent regulation, while emissions in agriculture will remain largely stable.

As Table 8.2 shows, unlike in the other case study countries, the transport sector is already the largest emitter of CO_2 emissions in France. Both sectoral contributions and total emissions have seen substantial changes since 1973. Overall, CO_2 emissions were reduced from 484 Mt to 367 Mt between 1973 and 1990, representing a reduction of nearly one quarter. Official projections estimate that if there had been no change in energy intensity and fuel use in electricity generation, emissions in 1990 would have been 697 Mt. The main reasons for this reduction were conservation measures and electricity being generated from nuclear power. Among OECD nations, only Sweden (see Chapter 10) has reduced its emissions more rapidly. However, virtually all the reductions in CO_2 emissions occurred before climate change became an issue and before the crucial 1990 deadline used by both the FCCC and the EU as a baseline year for reduction targets, which makes emission reductions difficult.

CO_2 emissions from the electricity industry represent less than 10 per cent of French emissions, in contrast to the 30–45 per cent in many industrialized countries, because of the large share of nuclear power. As Figure 8.1 shows, the shift to nuclear power over the past 15 years has been substantial. While in 1980 nuclear power accounted for only 26 per cent of electricity generation and coal and oil played a considerable role, by 1995 81 per cent of French electricity was generated from nuclear power, with a further 15 per cent from hydropower.

Table 8.2. *Changes in Energy Related CO_2 Emissions by Sector (1980–1993)*

Sector	1980	1993	Variation (%)
Transport*	95	132	+39
Domestic and tertiary	114	99	−13
Industry and agriculture	154	97	−37
Power stations	106	25	−76
Total**	469	352	−25

*Excluding vessels in international waters.
**Total for mainland France (excluding overseas departments and territories) with adjustment for climatic data.

Source: Premier Ministre, 1995

A further four nuclear reactors (5200 MW in total capacity) will be commissioned by 1998, but it is not expected that any new reactors will be ordered before 2000. Therefore, bearing in mind the seven year gap between the decision to construct a reactor and its start up, no further reactors are likely to come into service before at least 2005. While currently there is a clear overcapacity (which mainly serves to provide electricity exports, thus effectively benefiting the CO_2 balance of other EU Member States such as Italy and the UK), post 2000 the most imminent problem will be that of replacing old plant, as those reactors brought into service in the 1970s are coming to the end of their lives.

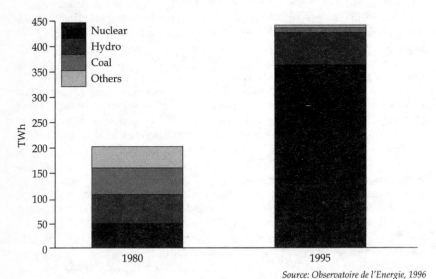

Source: Observatoire de l'Energie, 1996

Figure 8.1: Fuel shares in French electricity generation

Considering the already high penetration of CO_2-free sources in electricity generation, there is currently limited scope for further reductions in this sector. Fossil fuel plants are used as peak power plants and there is some scope for load management and flattening the power peaks, notably in the winter season, both by saving electricity and by displacing some kinds of consumption.

At present, the public monopoly company EdF accounts for almost all electricity generation in France, as well as for supply and distribution. An eventual liberalization of the electricity market, which will probably not exceed a vertical dismantling in France, and the introduction of some competition at the production level should manifest itself in cogeneration and gas power stations. However, this would occur to the detriment of coal and oil thermal stations, due to French techno-economic conditions, which means that nuclear power may stay cheaper than gas, although recent calculations based on low gas price scenarios demonstrate that gas may turn out to be cheaper. Therefore, on balance, CO_2 emissions should be reduced but only because of the economic advantage incurred with nuclear power given French energy policy.

FRANCE'S CLIMATE CHANGE POLICY

France's Environmental Credentials

It is fair to say that overall environmental awareness is less pronounced in France than in many Northern European countries. Green parties are less powerful, as illustrated by their winning less than 5 per cent of the vote in the 1995 presidential election. Their best electoral results occur in the European and regional elections, which in France are seen as far less important politically than are the presidential or parliamentary elections. Environmental themes, for example, were virtually absent from the agenda during the 1995 presidential campaign.

However, while not necessarily considered proactive in environmental policy, in certain areas the French government has been quite forward thinking. For example, in water resource management new unique institutions have been established. The so-called *Agences de Bassin* were among the first in Europe to introduce the polluter pays principle. Polluters are taxed and this money reverts to those who invest in antipollution measures. Another active policy area is that of the nature conservation. Special attention is paid to coastal areas: a public agency buys stretches of seashore to protect them from property development and other destruction. In matters of air pollution and solid waste disposal, French policy falls in line with European directives. France may be considered as the 'first follower', as compared to Germany as the 'first mover'. In some areas, however, French policy is active in taking initiatives and experimenting when Germany's proposals in Brussels are seen as undesirable. Thus, in response to the German *Duales System Deutschland* initiative on packaging waste, France developed its own *Eco-Emballage SA* to show that it was at least as effective, if not more so.

The Ministry for the Environment defines and coordinates environmental policy. But, in France, as elsewhere in the EU, a number of ministries are affected by environmental policy, including climate change prevention:

■ the Ministry of Industry, which is responsible in France for energy policy and controls the public energy companies;
■ the Ministry of the Economy and Finance, in particular via fiscal legislation and the macroeconomic effects of proposed measures;
■ the Ministry of Housing and Transport, in particular the *Direction des Transports Terrestres* and the section for housing standards;
■ the Ministry of Agriculture, for the management of soils and forest policy;
■ DATAR (*Délégation à l'amènagement du territoire et à l'action régionale*), the regional development agency.

Furthermore, since 1973, France has had a governmental agency – ADEME (*Agence de l'Environnement et de la Maitrise de L'Energie*) – responsible for energy and its effects on the environment, focusing in particular on the promotion of energy efficiency. This agency is under the auspices of both the Ministry for the Environment and the Ministry of Industry, akin to the UK's EEO (see Chapter 6). ADEME has local agencies in each region, which are responsible for initiating and supporting local activities, whether regional, departmental or communal. Considering its role in energy management and air pollution, ADEME is the obvious choice of agency to implement France's climate change policy and is playing an important role, as the subsequent sections show.

First Reactions to the Climate Change Issue

As already mentioned, to put France's attitude to the climate change issue in perspective, one has to keep in mind that the nation has already substantially reduced CO_2 emissions over the past 20 years as a result of an aggressive nuclear power programme and energy efficiency measures. Consequently, it is not surprising that the French government today, similar to Sweden (see Chapter 10), displays something of a laissez-faire attitude to the climate change issue. Discussions on which policy to adopt for climate change abatement did not begin in France until the end of 1991. Until this point, most of the actors who subsequently were to take part in this debate (industrialists, scientists, bureaucrats) were content to rely upon the lead that France held in this area – that is low levels of emissions per capita and per unit of GDP as compared with other OECD countries. It was generally held that, if anything needed to be done, it was that other countries, in particular France's main competitors Germany and the UK, should cut their emissions to French levels. This feeling was reinforced by the inaction of environmental groups and parties in France with regard to the greenhouse issue.

The EU Commission proposal for the carbon/energy tax, which was published in the autumn of 1991, had a catalytic effect. As a result, during 1992 and 1993 heated discussions took place. Studies were carried out which attempted to evaluate the economic and social effects of the measures proposed by the Commission (eg Giraud and Nadai, 1992; Beaumais and Brechet, 1992). The strongest opposition to the tax came from energy intensive industries which presented their arguments at the French and Community levels. The conflict subsided when the prospect of a Community tax became increasingly unlikely. A temporary agreement was reached on the basis that energy intensive consumers, whose competitiveness would be jeopardized in tax free zones, would be granted tax exemption subject to a 'voluntary agreement' to reduce emissions. However, industry still remains hostile to taxation.

In 1991, the Government set a target of limiting energy-related CO_2 emissions to below 2 t of carbon (or 7.3 t of CO_2) per capita per year by the year 2000, which with current population growth rates equates to a 14 per cent increase in total emissions by 2000. However, while this general commitment is still valid, the Government has declared that it should no longer be a specific target for 2000. Instead, it is very much in favour of target 'sharing' in the EU and the introduction of measures designed relative to a common reference of marginal cost of emission abatement (European Commission, 1996).

The actor that has so far played the most important role in developing French climate change policy is the Interministerial Committee on the Greenhouse Effect. The Committee was formed in the aftermath of the United Nations Conference on Environment and Development (UNCED) and was chaired from its induction in July 1992 to 1st of January 1995 by Yves Martin, a senior civil servant. It is now chaired by Pierre Chemillier. Martin coordinated the different public administrations involved and worked towards a consensus point of view to be presented by France at EU discussions. It elaborated a programme in accordance with the FCCC requirements which was published in February 1995, under the auspices of the Prime Minister, as the *Programme National du Prévention du Changement de Climat* (the National Programme for Climate Change) (Premier Ministre, 1995, summary in ADEME, 1995).

The committee's approach was to try to identify the marginal costs of reducing emissions in each greenhouse gas producing process. Then it considered in greater detail the possibility of reducing emissions in processes where the cost was less than 70 ECU per tonne of reduced carbon, and tried to find regulatory measures or incentives to instigate these reductions effectively. Evidently this method was primarily theoretical, and was often hampered by the lack of data on the cost of reducing emissions. Furthermore, different ministries, often influenced by different interest groups, were attempting to push their own agendas.

The Interministerial Committee gradually managed to reconcile the diverging points of view, and the National Programme can be considered as representing a consensus position. Conversely, what has fluctuated in the past, and will continue to do so, is the extent to which the government is prepared to push the issue at EU and international levels. At the time of writing it is impossible to predict the attitude of the Minister for the Environment, Corinne Lepage, and the Juppé government in general towards this issue. It seems that the French government is waiting for the 1997 meeting of the FCCC in Kyoto to define more precisely its position and proposals.

The French National Programme

In the National Programme, previous achievements in terms of CO_2 reduction and energy savings are stressed. The Programme effectively relies on a continuation of previous energy efficiency policies, plus a small number of specific climate change measures. There has been some recognition that, on account of the current low energy prices, policies in the energy efficiency area are likely to be less effective now than in the past. However, while in principle in favour of fiscal measures to amend this situation, the Government insists that these measures must be taken at the EU level and that they are to be carbon specific, as opposed to the combined carbon/energy tax which would have penalized nuclear power.

As Table 8.3 shows in some detail, the Programme consists mainly of regulatory measures, fiscal incentives for energy saving, improved infrastructure and the support of R&D. While the expected CO_2 emission savings of a number of individual measures are specified, the information given is not sufficient to derive a quantitative estimation of the global effect of the set of measures. Furthermore, the implementation of various measures is currently uncertain.

ENERGY MEASURES

Electricity Generation

Security of supply has been the main focus of French energy policy for more than 20 years and, in view of France's lack of indigenous fossil fuel reserves, this policy aim has also resulted in an incidental and substantial reduction in CO_2 emissions. Since the oil crisis in 1973, France's energy policy has been composed of two parts:

■ A vigorous energy efficiency policy based on regulatory measures, particularly with regard to new buildings, on incentives, such as the cofinancing by ADEME of R&D programmes and fiscal incentives to invest in energy efficiency in industry and construction, and on high taxation on fuel.
■ A policy of rapid development of nuclear power.

Environmental groups protested against the nuclear option in the 1970s, but this protest weakened during the 1980s and there is currently broad consensus, despite some minority opposition groups. In terms of CO_2 emissions, as already discussed, the switch to nuclear power in electricity generation has been particularly significant. However, there have been some negative implications as well, so that there is still some scope for emission reduction from electricity generation. The first is linked to the overcapacity of nuclear power and the resulting promotion of electrical home heating. This increase has created high demand peaks in winter, which are met by thermal power stations using coal or oil. The CO_2 balance when compared with mains gas heating is most probably negative but some effort is being made to tackle this problem. EdF has drawn up an adjusted tariff rate and has signed an agreement with ADEME to curb demand when supply produces CO_2 emissions (for example in overseas departments and Corsica). The government has asked EDF to set aside FF 100 million per year for this task.

The second undesirable effect has been caused by EdF's monopoly position. Combined with the nuclear overcapacity, this has led to rural electrification, notably in the number of cables laid for uses, such as heating, which do not necessarily require electricity, and which are unjustified from an economic as well as environmental point of view. Furthermore, this trend threatens the possible development of renewable energy in isolated rural areas. Some limited but positive measures to counter these effects have been included in the National Programme.

The French government has to date been very negative about the prospects of liberalization in the energy sector, as is demonstrated by its attitude to the EU Internal Energy Market proposals. However, although privatization of EDF seems unlikely, vertical disintegration within the company and the stimulation of competition at the production level (with the entry of private producers) and at the distribution level will probably be introduced. This should reinforce the mechanisms which offset the perverse effects of an overly centralized electricity network. Whatever their form, these effects should not be given too great an importance in comparison to the massive reductions in CO_2 emissions gained from the nuclear programme.

Energy Efficiency Policies: Some New Measures

Some of the problems of the current energy system have been recognized by the government which in 1994 held a national debate on ways to improve security of supply, environmental compatibility and energy efficiency. After a series of public meetings and symposia involving more than 6000 people, the *Souviron Report* (named after the rapporteur of the initiative) recommended as one of its main conclusions that the commitment to solving the CO_2 problem would require more efforts in the energy efficiency field (Souviron, 1994; Conseil des Ministres, 1995).

On the whole, the primary policy aim of improving energy efficiency has been achieved successfully, with a considerable reduction in energy intensity (at an average of 0.5 per cent per annum between 1979 and 1992). However, in recent years some of the earlier efficiency gains have been lost. Energy saving activities have decreased since the fall in oil prices, from the mid 1980s onwards, resulting in a worsening of the efficient of energy use, mainly taking place from 1991 onwards. Additionally, a series of other obstacles have manifested themselves. Firstly, actors (industries as well as households) do not display economic rationality; they do not make all the

Table 8.3. *Main CO_2 Reduction Measures in the National Programme*

Area of implementation	Measure	Objective	Expected emission reductions	Implementation as of June 1996
New residential dwellings	New building standard in 1997	10% improvement in energy efficiency	0.6 Mt CO_2 p.a. by 2000	Still under discussion
New service sector	Energy labelling			Not yet decided
	New building standards in 1997/1999	25% improvement in energy efficiency	4 Mt CO_2 p.a. by 2000	
	Labelling, information			
Existing buildings	Financial assistance for energy audits		12 Mt CO_2 p.a. by 2000	Only for the principal residence
	Direct and fiscal incentives for investment			
	Promoting increased usage of wood in construction	Triple use of wood between 1990 and 2000	2.6 Mt CO_2 p.a.	Not yet
Industry	Voluntary agreements	Promote investments with 4 or 6 year pay-back		One agreement signed
	Fiscal incentives for energy efficiency and labelling of materials			Not yet
Electricity sector	Promotion of DSM			Yes, but downsized
Road freight	Technical standards for vehicles		1.2 Mt CO_2 p.a.	Not yet
	Increased competition		1.5 Mt CO_2 p.a.	
	Development of combined road-rail networks	Double by 2000		Yes
Cars and light vehicles	Technical improvements (R&D)	FF 1.2 billion of subsidies between 1990 and 1994		Yes

	Subsidized scrapping of old vehicles	Scrapping of 400,000 vehicles over 10 years old	Yes
	R & D and testing of electric urban transports		Minor
	Development of urban transport systems	FF 1.2 billion p.a. in public subsidies. FF 5 billion p.a. in investment by local authorities	
Renewable energy	Development of TGV network	+700 km of track by 2000.	Scheduled
	Promotion of renewable energy sources in rural areas	Subsidy of FF 100 million from 1993–1995	Yes
	Development of wood-based community heating systems		Yes
	R & D, testing and tax exemptions for biofuels	Ethanol production to rise from 27,000 t in 1993 to 110,000 t in 2000; rape seed oil from 8000 t to 400,000 t in 2000	Yes
	Waste incineration	Triple between 1990 and 2003 – Savings: 1 Mtoe	Yes
Forests and soil	Subsidized reforestation	Increase of 10,000 hectares per year to 30,000 hectares per year in 1998	Yes; 9 Mt CO_2 accumulated between 1990 and 2000; 300 Mt CO2 in 50 years

Soure: Premier Ministre, 1995; and author's own data

investments in energy efficiency that would be profitable for them. This requires an information and labelling policy to improve the rationality of choices. Secondly, and more importantly, the cost of energy and the uncertainty of prices in the future present major obstacles. From this perspective, a strong commitment by the EU to instigate a clear fiscal policy, gradually augmenting the price of energy in relation to CO_2 emissions, is favoured by the French government as a powerful signal to re-initiate long term efforts towards increased energy efficiency.

While there clearly have been previous efforts in France to improve energy efficiency, some doubts can be raised as to the French government's commitment to energy efficiency since ADEME's budget was cut from FF 483 million in 1992 to FF 283 million in 1994 and has been at this low level since (FF 270 million in 1996). The National Programme does include a number of measures to improve efficiency. In the domestic sector, new building standards will be applied as of 1st January 1997, leading to a 5–10 per cent reduction of heating requirements in new buildings. For new tertiary sector buildings, in order to encourage professionals to build more energy efficient structures in the future, seven sector guides have been published (hotels, offices, health and education, trade, leisure, industry and agriculture). Furthermore, a new building standard which aims to increase energy efficiency by 25 per cent will be applicable from 1st July 1997 for buildings that are not air conditioned and from 1st January 1999 for those that are air conditioned.

For existing buildings there are a number of grants and subsidies available, mainly through ADEME. Investments in energy efficiency amounted to around FF 39 billion in 1994. Furthermore, there still exists ample scope for further improvements. EdF has only recently become involved in DSM. An accord signed by EdF and ADEME in 1993 shows that 19 regional and three national DSM programmes were launched, covering the promotion of efficient lighting and appliances, energy efficiency audits in industry and efficient industrial motors. However, spending by EdF, ADEME and the local authorities on these programmes has been less than planned (in 1994–1995 only FF 10 million instead of the planned FF 24 million).

As far as industry is concerned, since 1973 CO_2 emissions in this sector have been reduced by 38 per cent, mainly through a reduction in the marginal emissions of production, but also through changes in industrial structure. More than three quarters of industrial CO_2 emissions are produced by a few energy intensive sectors, comprising around 1000 firms: aluminium, iron and steel, construction materials, refining, cardboard manufacture, glass, food industries and chemicals. There is still scope for energy efficiency improvements, but industry has been very vocal in its opposition to proposals for taxation, and has threatened relocation.

As a result, the Government has decided to pursue voluntary agreements within industry, in return for exemptions from an eventual tax. ADEME studies have identified potentials for a 20 per cent reduction in CO_2 emissions through energy saving or fuel substitutes, provided industrialists are prepared to accept a longer time period on investment returns than is usual in industry (CEREN, 1994). Discussions with representatives from the energy intensive industries (chemicals, iron and steel, cement, construction materials) did take place but failed, essentially because of the collapse of the carbon/energy tax proposals at EU level. As industry no longer felt threatened by the tax it saw no need to enter into an agreement. Meanwhile there are a number of instruments which should ensure some emission reductions from this sector:

■ An obligation, currently under amendment, to perform energy audits in industries that consume more than 300 toe per year.
■ Financial aid: R&D support, consultancy advice and grants for prototype investment (total: FF 398 million from 1990–1993, downsized to FF 31 million in 1996).
■ Tax incentives for energy saving materials and cogeneration: quick amortization and 50 per cent or 100 per cent reduction of professional tax base for equipment, tax exemption on profits and property value increases produced from loans by *Sofergies* (*Sociétés agrées pour le financement des économies d'énergie*).

Overall, France has a reasonably active energy efficiency policy with a range of measures. However, these are unlikely to be sufficient to yield substantial savings in the current climate of low energy prices, especially within the industrial sector (see also IEA, 1996). Overcapacity in the electricity sector has meant that EdF has been somewhat reluctant to become involved in DSM or to support the development of CHP. The lack of a favourable environment for CHP was highlighted in the *Souviron Report* and in February 1996 a new purchasing tariff for cogenerated electricity was introduced. In CO_2 terms, it has to be kept in mind that CHP and improvements in electricity efficiency (eg lighting, appliances) will only lead to marginal emission reductions in France, due to the already low carbon intensity of electricity generation.

Support for Renewable Energies: The Cinderella Treatment

The last point above is also valid for renewable energies. Nevertheless, the development of these energy sources has a range of other benefits and is crucial in the move towards a more sustainable energy system. Renewable energy has certainly suffered in the shadow of the nuclear power expansion, as the large overcapacity in nuclear plant has meant little interest in other electricity generating options. Nevertheless, with an installed capacity of some 26 GWe, France has the largest and one of the most exploited hydropower resources of the EU Member States (European Commission, 1994) and fuelwood supplies nearly one quarter of domestic heating demand. Furthermore, the La Rance tidal barrage is the only large scale development of tidal energy worldwide. Also, France is virtually alone in having experimented with solar thermal electricity generation, although the two pilot plants have since been closed on commercial grounds. Finally, geothermal resources have also been developed. While France already has a reasonable utilization of renewable energies, strong barriers exist to their further development, in particular the rules for grid access and limits on the size of plant for independently generated electricity set by EdF. Buy back rates are reported among the lowest in the EU (European Commission, 1994).

There are some government measures aimed at increasing the contribution of renewable energies, principally targeting biomass technology and waste incineration. A 'wood energy' plan has been drawn up by which the government and local authorities in a number of pilot regions will create the infrastructure for wood based community heating systems, in terms of both supply and demand. The management of the programme has been entrusted to ADEME. After wood, agricultural biomass is the main focus for the development of renewable energy, in particular biofuels to be used in the transport sector. Substantial increases in the production of ethanol and rape seed oil are planned (see Table 8.3). However, the IEA (1996) is highly critical of this emphasis on biofuels, which according to several government-commissioned studies is dubious, both in terms of long-term cost-effectiveness and in terms of environmental benefits.

There has been limited support for solar and wind power, the potential of which is seen as very restricted, except in certain isolated rural areas. The development of small hydro-electric stations, while possible, is not likely for environmental reasons, such as the preservation of natural beauty and of wildlife. No funding has yet gone to wave power, although France is reported to have the largest potential (other than the UK) in the EU (European Commission, 1994). An agreement has been drawn up between ADEME and EdF for the development of renewable energy and DSM in rural areas, but this is currently hamstrung by the harmonization of tariffs. An annual budget of FF 100 million, liable to be increased, has been allocated for 1993–1995 to finance measures taken within the framework of this agreement. The first wind turbines under this scheme are in operation, and others are under construction.

At the time of writing the prospects for wind power are beginning to look much better. In February 1996 the French industry minister launched a new programme for wind power, aimed at reaching a capacity of 250–500 MW by 2005, with joint calls for tenders from ADEME and EdF. Special procurement conditions for electricity generated by wind power will be introduced with a view to reaching commercial viability within 10 years. Additional preferential conditions will apply to the French overseas departments and Corsica. Furthermore, EdF intends to increase its spending on wind and solar energy to FF 125 million over the next five years, up from the 1995 FF 4 million. It remains to be seen whether this programme will be more successful than previous ones, but it is certainly a step in the right direction.

TRANSPORT POLICIES

Upward Trends

In view of the low emissions from the energy sector, the transport sector is even more fundamental to future emission cuts in France than elsewhere in the EU. However, recent trends give little hope for optimism. Between 1980 and 1993, CO_2 emissions from the transport sector increased by 39 per cent, with no sign of slowing down since 1993. This increase has been the result first and foremost of the sheer increase in passenger traffic volumes, but also in a shift from public transport to the private car. As Table 8.4 shows, in freight transport the increase has been less pronounced although there has been a substantial increase in lorry traffic.

Table 8.4. *Developments in Traffic Volumes*

Base 100 = 1970	Passenger transport:	190
Indices in 1990	Aeroplane	274
	Car	194
	Bus + metro	154
	Train	157
	Freight transport:	120
	Lorry	198
	Train	75

Source: Premier Ministre, 1995

Between 1967 and 1986 the average distance from house to working place rose from 6 km to 11 km (Premier Ministre, 1995). Most of this increase is due to the use of private vehicles. The increase in road traffic has taken place despite France having a policy of high fuel taxation. Taxes currently account for over 80 per cent of the final price of a litre of petrol, which is the highest rate in the whole of the OECD. Petrol prices themselves are the third highest after Norway and the Netherlands (IEA, 1996). Meanwhile, the development of the high speed TGV (*Train à grande vitesse*) train, which competes with aeroplanes and private vehicles, has brought about a clear reduction in energy consumption and an even greater reduction in CO_2 emissions. As such, France has less of a road bias in its transport policy than a number of other countries.

Yet, transport emission trends are worrying and there is little indication of the government seriously tackling this issue. To some commentators the fact that the Infrastructure Minister presented a state financial package for 1500 km of new motorways the same day as the Environment Ministry presented the National Programme for climate change is indicative of a lack of integration of climate change concerns into French transport policy (Climate Action Network, 1995).

Passenger Transport

As far as passenger transport measures are concerned, the National Programme is somewhat curious in that it places much emphasis on government subsidies for buying new cars. This involves a subsidy of FF 5000 (for small cars) or FF 7000 (for large cars), provided the buyer scraps a vehicle older than 10 years. However, this subsidy has not been introduced for CO_2

emission reduction reasons but to aid the French economy, as most French people will buy cars from French manufacturers. While there may be some CO_2 emission benefit, this assumes that new cars are more fuel efficient than old cars, which is not necessarily the case. Scholl et al (1996) have shown that the energy intensity of cars (measured in MJ per vehicle km) only decreased by 4 per cent in the period 1973 to 1992, while the carbon intensity per passenger km actually increased by 1 per cent. Clearly, if the government had been aiming at a CO_2 emission abatement through this measure it should have related the subsidy to fuel consumption criteria, which would mean that, generally, larger cars should receive subsidy than smaller cars. Moreover, nothing is done to deter car buyers from buying air conditioned cars, although air conditioning increases fuel consumption in urban traffic by up to 80 per cent per km.

The Government is in favour (at least according to the National Programme) of consumption standards for cars, with possibly a system of negotiable permits, but sees the need for EU action rather than national action. Meanwhile, FF 1.2 billion were made available between 1990 and 1994 for R&D on more efficient cars. Furthermore, efforts have been made to develop and promote electric vehicles, which have a particularly low CO_2 emission rate in France. It is also worth noticing that the new *Loi sur l'air* (Air Pollution Act) issued in 1996 obliges urban authorities to draw up Urban Transportation Plans, increase the cost of parking, provide fiscal incentives to use electric or gas powered cars in cities and modify the rules of urban planning in order to encourage public transportation.

As far as public transport is concerned, reliance will be on further development of the TGV network, which has high energy efficiency and replaces fossil fuel with electrical power. At the end of 1994, the network covered 1260 km (Southeast and Northern Europe, the Atlantic coast and cross Channel). Before 2000, the *TGV Méditerranée* will be extended to Marseille and the *TGV Est* will be brought into service, adding a further 700 km of track. Action by local authorities is also vital in this sector. The government subsidizes investments in urban transport modernization and, since 1994, has tried to encourage the promotion of public transport. From 1989 to 1993, FF 1.3 billion were invested annually; this will be sustained until 1998, increasing annual investment to approximately FF 5 billion.

Freight Transport

The fast growth of 4 per cent per year in freight transport is primarily due to long-distance haulage and international transportation (+6.7 per cent per year) (Premier Ministre, 1995). This is explained partly by the expansion of the EU to include Spain and Portugal, and subsequently the creation of the Single Market. The tariff deregulation introduced in France in 1986 has heightened competition, which in turn has resulted in an increased incidence of rule breaking, such as speed limits, weight and driving time limits. Economies thus made on the part of road hauliers have contributed to lowering road transport costs (–3.4 per cent per year between 1985 and 1992). The market share of road haulage increased subsequently from 58 per cent to 69 per cent in the same time period. In coming years, an average annual increase of about 2.5 per cent for internal traffic is predicted (5 per cent for foreign heavy goods in transit). As with passenger transport, this trend seems little affected by high fuel taxes. Excise duty on diesel is currently 31 per cent higher in France than the minimum EU rate. France is proposing further increases, but only in conjunction with other EU Member States.

One of the aims of French transport policy is to develop alternatives to road based freight transport, notably long distance road–rail networks, especially on routes with high density traffic or through difficult terrain, such as the Alps. To this end, financial aid is given to R&D in the combined road–rail transport sector (FF 450 million from 1990 to 1994). New transfer yards will be constructed before 2000 in Bordeaux, Marseille and Lyon. Studies on the possibility of a TGV line from Lyon to Turin will include provision for freight transport. The objective for 2000 is a doubling of combined transport since 1990.

In general, all measures which have been taken or, indeed, could be taken to encourage the use of public transport or the intermodal transport of goods, for example, are secondary to the central problem of the growing mobility of individuals which is currently subsidized. The use of cars in towns is effectively subsidized by the failure to make drivers pay the total external costs that this usage incurs. The growing dispersed urbanization of areas around towns also makes the use of a car essential, with out-of-town shopping centres common everywhere. For this reason, transport policy in France, as in many other countries, cannot be said to be sustainable. The measures proposed in the National Programme are certainly useful, but they do not attack the fundamental problem of the growing need for car usage, especially in towns. At the same time, it has to be kept in mind that France is a large country with a relatively low population density, thus some long distance travel is a necessity for communication between different parts of the country. Also, France is a transit country for much intra-European freight traffic and there is limited scope for the country to affect emissions without a wider EU framework in this area.

LOCAL AND REGIONAL AUTHORITY INITIATIVES

France remains a very centralized country, which leaves little political room for manoeuvre at the regional and municipal levels in matters of greenhouse gas emission reduction. In particular, in the energy sector regional and local authorities have almost no power, very much akin to the situation in the UK. Some local authorities, such as Rennes, run their own district heating systems but otherwise have no involvement in the energy sector, which is dominated by public monopolies. However, there are at least three areas of differing importance where subnational authorities, and in particular those at the municipal level, can play a role:

■ Influencing urban car use through the development of public transport and use of planning controls.

■ Improvements in energy efficiency in public buildings.
■ The promotion of renewable energy.

Local initiatives do exist, although usually within the framework of more general local environmental policies, rather than of specific local climate change policies, as found in Germany (see Chapter 5). Towns, with the help of ADEME, have previously been the advocates of geothermal programmes for town heating. These programmes are now inactive as, due to the fall in fossil fuel prices, they are no longer profitable. The programme for the use of wood in construction is also being developed at the local level, with the help of ADEME. Furthermore, in 1990 ADEME launched a programme entitled *Citévie*, aimed at developing local energy management policies. A total of 25 towns and cities have participated in the programme, with ADEME funding used to carry out studies and energy audits (IFEN, 1994). Pilot schemes for the development of electric car pools have been implemented in some towns, such as La Rochelle. Finally, towns are responsible for the disposal of municipal waste and thus for recycling initiatives or the use of waste incineration for power generation.

Some towns have been very active in improving energy efficiency in public buildings. State authorization to employ private companies that offer energy efficiency services has facilitated action in this area. Rennes, for example, has reduced energy consumption in municipal buildings by 50 per cent. Furthermore, solar water heating systems have been installed at several municipal buildings and the city authorities have run various public awareness campaigns on energy efficiency and renewable energies (OECD, 1994).

The most important areas of influence are the management of urban traffic and the development of public transport systems. Numerous towns have prioritized the latter in their programmes, manifesting themselves in such ways as the reappearance of trams in large provincial towns, such as Strasbourg, and the development of metro systems in Marseille and Lyon. However, the financing of such initiatives is problematic. It would seem logical to fund these schemes by taxing some of the external costs incurred by urban car use. Unfortunately, municipal councils are not currently empowered to do this. Curiously, although this would seem to be a matter for the national government, being a fiscal reform, the government itself claims that an urban car tax would be a European matter. Taxing urban car use is also seen as increasing social inequalities.

Overall, local and regional specific action on climate change is rather limited in France. The issue has not really interested local authorities, who generally are somewhat behind other countries, such as Germany or the Netherlands, in the development of local environmental policies. Nevertheless, efforts to develop public transport systems have substantial CO_2 benefits. Some towns and regions have participated in the various EU programmes on local and regional energy management, but French towns and cities have not been active in ICLEI or the Climate Alliance.

CONCLUSIONS

As this chapter shows, France's response to the climate change issue has been shaped fundamentally by the dominance of nuclear power in the energy sector. Hence, France is not in favour of response strategies which are based on flat emission reduction targets, arguing that these are very difficult to achieve due to the disparity of previous efforts and national circumstances. Instead, France favours policies which specify the means by which all developed nations may reduce all emissions below a common threshold cost. In keeping with this attitude, France has not been very concerned about achieving the EU stabilization target for 2000 and its national programme relies principally on the continuation of previous policies aimed at cheaper and more self-sufficient energy production, and at certain measures that provide multiple benefits, for example reforestation as prescribed by the reformed Common Agricultural Policy or the management of waste disposal.

The French climate change programme is characterized by regulations, financial incentives and public infrastructure construction. Consequently, the main obstacle to the effective implementation of such a programme is the possibility of a political U-turn which would seriously reduce its effectiveness, not least by the reduction of financial aid. The likelihood of stabilizing emissions in 2000 at the level of 1990 depends largely on the independent policy factors already mentioned, namely economic growth, fossil fuel prices and the availability of nuclear power. These factors may affect the predictions for total emissions in 2000 by ± 8 per cent.

In general, the current programme is insufficient to provide a long term stabilization of emissions, let alone a reduction. Analyses show that, even with the above measures, a growing 'hard core' of CO_2 emissions (around 4 per cent per decade) is left untouched (Commissariat Général du Plan, 1995) resulting mainly from the fast growth of transport emissions. In this sector, current policy measures are clearly inadequate and other policy decisions (investments in motorways) demonstrate a lack of consideration of CO_2 factors in transport policy. In essence, this reflects developments in the EU as a whole.

To proceed further the French government sees the need for a common EU strategy, using economic instruments such as taxation and emission permits, and coordinated with Europe's global partners. Generally, climate change is not seen as a priority in French environmental and general economic policy making. This is not entirely surprising, given that public opinion in France shows little concern for the climate change issue, with a general feeling that a small country like France, or even the EU, can do little about such a global problem. Moreover, not only public opinion but also many policy makers are not really informed about the climate change issue. For example, almost everybody thinks that the temperature may increase in Western Europe, while recent studies show that, due to the disappearance of the Gulf Stream, the average temperature may drop by up to 4°C. This attitude of little concern is not condemned by the environmental movement. Except for certain individuals speaking on their own behalf, the environmental movement in France has said little about the issue. As a movement, it is far more capable of mobilizing support and action against local pollution. Without much public concern, there is little political will to act, although a number of measures (especially more vigorous energy efficiency improvements) would have general economic benefits.

REFERENCES

ADEME (1995) La France et l'Effet de Serre, ADEME, Paris

Beaumais, O and Brechet, T (1992) *Investissements Économiseurs d'Énergie et Écotaxe: Une Mise en Perspective Macro-économique*, ERASME, Ecole Centrale de Paris et Paris–1

CEREN (1994) *Evaluation de l'Impact du Gisement d'Actions de Maîtrise de l'Énergie dans l'Industrie sur les Émissions de CO$_2$* (Etude pour ADEME, confidential), CEREN, Paris

Climate Action Network (1995) *National Plans for Climate Change Mitigation: Independent Evaluations*, Brussels

Commissariat Général du Plan (1995) *Energie 2010*, La Documentation Française, Paris

Conseil des Ministres (1995) *Communication des Ministres Chargés de l'Energie, de l'Environnement et de la Recherche sur l'Energie et l'Environnement*, 29th March 1995, Paris

EdF (1996) 'Electricity generation 1995', as quoted in *Power in Europe*, 3rd March 1996, p3

European Commission (1994) *The Renewable Energy Study, Annex 2, Country Profiles*, DG XII, Brussels

European Commission (1996) 'Second evaluation of national programmes under the monitoring mechanism of Community CO$_2$ and other greenhouse gas emissions', *COM* (96) 91

Giraud, P N and Nadai, A (1992) *Taxation des Émissions de CO$_2$ et Compétitivité de l'Industrie en France*, CERNA, Ecole des Mines de Paris

IEA (1996) *Energy Policies of IEA Countries. France 1996 Review*, OECD/IEA, Paris

IFEN (1994) *L'Environnement en France (Edition 1994–1995)*, Dunod, Paris

Observatoire de l'Energie (1996) *Bilan Énergétique Provisoire de l'Année 1995*, Ministère de l'Industrie, Paris

OECD (1994) *Urban Energy Management: A Handbook of Good Local Practices*, OECD, Paris

Premier Ministre (1995) *Programme National de Prévention du Changement de Climat*, Office du Premier Ministre, Paris

Scholl, L, Schipper, L and Kiang, N (1996) 'CO$_2$ emissions from passenger transport', *Energy Policy*, Vol 24, no 1, pp17–30

Souviron, J P (1994) *Débat National sur l'Energie et l'Environnement, Rapport de Synthèse*, ADEME, Paris

Chapter 9 | SPAIN: FAST GROWTH IN CO$_2$ EMISSIONS

Xavier Labandeira Villot[1]

INTRODUCTION

The possibility of human-induced climate change processes has caused considerable public concern in Spain during the past few years. Persistent climate extremes, with large areas in southern Spain suffering drought and heat waves, actions by the environmental movement and increasing media attention have raised the issue to a pre-eminent position. In fact, Spain may suffer more from climate change than any other EU Member State. Climate scenarios for Spain indicate that a doubling of CO$_2$ concentrations in the atmosphere may lead to large increases in temperature (of around 2.5°C in the annual average), generalized decreases in rain and soil humidity and greater variability between and within years. These phenomena will be felt particularly in central and southern Spain. Other worrying effects include an increase in the frequency of severe storms in the Mediterranean and a general sea level rise that would affect the Spanish coastline and its multiple islands. Also, three of the most current pressing environmental problems, desert advance, soil erosion and water scarcity, are likely to worsen due to global warming.

The former Spanish government reacted to domestic pressures and the process of international commitment by stating that climate change constituted one of its main concerns and the most serious challenge for its environmental performance. Nevertheless, there is no real climate change policy at the moment in Spain, although a climate change strategy is being developed. It is true that there are public policies in place that may have effects on climate change processes, but they have been designed with other objectives in mind.

At the time of writing (June 1996), the new government has been in

1 The author gratefully acknowledges the assistance of all those officials who gave their time and insight during the interviews carried out for this study.

office for just a few weeks, so it is not yet possible to offer an assessment of its actions or projects. However, an analysis of its manifesto (Partido Popular, 1996) indicates that there will be few variations from the strategies and objectives pursued so far on these issues, the most significant change being the creation of a separate Ministry for the Environment.

In this chapter an attempt is made to provide an analysis of the opportunities and constraints for the introduction and development of effective climate change policies in Spain. A description of the current level and sources of greenhouse gas emissions, focusing on CO_2, is followed by a presentation of current public policies that are relevant for climate change. Finally, comment is made on the possibilities for action in the future based on the analysis and comments from different public sector departments and individuals.

SPANISH GREENHOUSE GAS EMISSIONS

Spain is the EU's fifth largest CO_2 emitter, and hence one of the Member States particularly important for emission reductions. A serious and comprehensive assessment of Spanish greenhouse gas emissions has been carried out only relatively recently. The 1990 inventory of emissions using CORINE-AIR methodology was the first attempt and the most up-to-date figures, from 1993, are shown in Table 9.1.

Table 9.1. *1993 Inventory of Spanish CO_2 Emissions*

Emitting sector	Emissions (ktonnes)	Emissions (%)
Electricity generation	65,790	24.4
Nonindustrial combustion plants	29,481	10.9
Industrial combustion processes and plants	64,432	23.9
Industrial processes without combustion – total	15,611	5.8
Cement	11,105	4.1
Transport – total	63,480	23.5
Road transport	50,206	18.6
Urban car use	13,878	5.1
Buses and lorries	13,444	5.0
Others	31,198	11.5
Total	269,992	100.0

Source: MOPTMA, 1994

Table 9.1 indicates that Spanish CO_2 emissions largely come from a few activities and economic sectors. Electricity generation alone accounts for one quarter of all emissions, roughly the same amount as transport. Another 20 per cent is emitted both by the industrial sector and by residential and industrial combustion plants. In short, more than two thirds of Spanish CO$_2$ emissions come from energy related processes.

As far as CH$_4$ emissions are concerned, decreasing coal production will reduce some of the emissions related to coal mining (20.6 per cent of the total). Most of the remainder is related to waste disposal sites (19.3 per cent), cattle farming (26.4 per cent) and forests (21.4 per cent). Economic development will encourage waste production with a probable increase of emissions from waste disposal sites. Agriculture and forestry are also the largest culprits of N$_2$O emissions. In these areas, the possibilities for N$_2$O and CH$_4$ abatement seem to be limited. Finally, there is a clear trend towards increasing emissions of the precursors of tropospheric ozone (O$_3$), another greenhouse gas, as a result of transport expansion.

Figure 9.1 shows the evolution of Spanish energy related CO$_2$ emissions during recent decades. According to OECD data there has been a clear growth in emissions, with an increase in the absolute and relative weight emissions of the transport sector and a simultaneous decrease of emissions from industrial activities. The rise in CO$_2$ emissions was especially significant between 1988 and 1992 due to the extensive growth in the Spanish economy. However, CO$_2$ emissions decreased in 1993 for the first time since the early 1980s due to the harsh recession (Figure 9.2).

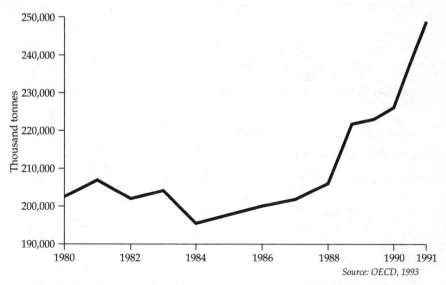

Source: OECD, 1993

Figure 9.1: Energy-related Spanish emissions of CO$_2$ (1980–1992)

In terms of per capita emissions, Spain, with 5.69 t CO$_2$ per capita, lies well below the EU average of 8.46 t. This low figure is the result of two factors – a warm climate and a low level of economic development. However, the gap between Spain and the EU average is decreasing. In fact, the EU as a whole seems to have been stabilizing per capita emissions since the 1970s, while Spain has observed a continuous and strong increase during the past 30 years.

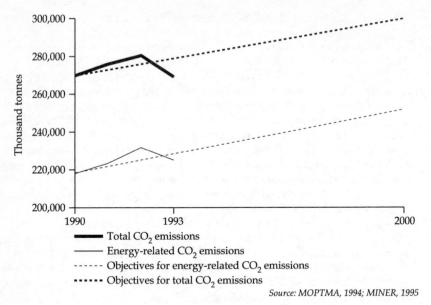

Source: MOPTMA, 1994; MINER, 1995

Figure 9.2: Spanish total and energy-related emissions of CO_2 (1990–2000)

Figure 9.2 gives a projection of future trends in Spanish CO_2 emissions. Included here is information provided by the recent CORINE-AIR inventories of emissions and the official forecasts set in the Strategy for Energy and the Environment, ESEMA (MINER, 1995). The National Energy Plan 1991–2000 (Ministry of Industry, 1991), *PEN 91*, had forecast a 25 per cent increase in CO_2 emissions from the energy sector by the year 2000 with respect to 1990 levels. The projections in Figure 9.2 imply an emissions increase of 20 per cent less than those in *PEN 91*, mostly due to the introduction of a programme to encourage energy efficiency. However, the recession between 1992 and 1994 and political pressures on the government have led to a review of the ESEMA. The revised version projects only a 15 per cent increase in energy related emissions by 2000.

In terms of the future development of CO_2 emissions, it is important to recognize a number of broad features that characterize the Spanish energy system:

■ There is less effort put into energy efficiency measures than in most EU countries, with unfavourable energy intensity ratios (final energy/GDP).
■ Spanish fossil fuels have poorer environmental and energy qualities than those imported.
■ There is high dependence on fossil fuels for electricity generation, with decreasing hydro generation due to severe drought, and stabilization of nuclear generation after the 1984 moratorium.

■ Natural gas is in the process of attaining a substantial position within the Spanish energy system.
■ There seems to be little use of price policies to achieve energy consumption reductions, owing to the rigidity of the market.

The evolution of Spanish consumption of primary and final energies is clearly related to economic performance and thus shows a remarkable rise in the late 1970s and late 1980s and the opposite in the early 1980s and since 1992. There is an increasing gap between primary and final energy consumption, a symptom of the inefficiencies of the Spanish energy sector. The values of two indicators of energy efficiency, the relationship between primary and final energy and the consumption of energy per unit of GDP, show the possibilities for efficiency improvements in Spain. These have followed a path of negative evolution since the early 1990s and do not perform well in comparison to the European averages.

Electricity generation in 1994 was dominated by coal (39 per cent) and nuclear power (34.4 per cent), with hydropower supplying another 17.5 per cent (IEA, 1996). Since the oil price shocks of the 1970s, Spain has embarked on a large oil substitution programme in electricity generation, focusing on coal and nuclear power. However, because of the nuclear moratorium the role of nuclear power is to decline over the coming years (eg to 23 per cent by 2000), with natural gas expected to make up most of the shortfall. This will obviously have negative implications in CO_2 terms, even if gas is burned in efficient power stations and CHP plants. On the positive side, the proportion of coal fired production is expected to decline to around 34 per cent by 2000. Meanwhile, hydropower is seen to have reached its maximum potential and may indeed decline if rainfall patterns continue as in recent years. The role of other renewable energies is set to increase (see p156).

Overall, Spanish CO_2 emissions are on a clear upward trend but their exact growth rate will depend on levels of economic growth, fuel use in the electricity sector and energy efficiency developments.

THE NATIONAL PROGRAMME FOR CLIMATE CHANGE

Along with the rest of Southern Europe, Spain is usually considered one of the laggard countries in environmental policy. Undoubtedly, membership of the EU has provided a great spur to the development of Spanish environmental policy, but Spain has repeatedly been one of the countries with the worst implementation record for EU environmental directives (Aquilar, 1993). To some extent this is excusable. Spain, as a relative latecomer to the EU with its accession in 1986, has had to take on board a large body of EU environmental legislation without having been able to influence its development. In addition, according to Aquilar (1993), there is a feeling of resentment against an environmental policy that is seen as chiefly a response to the ecological problems of the Northern Member States.

However, as already mentioned, the climate change issue has clear negative implications for Spain itself and in recent years there have been signs of environmental issues being taken more seriously in Spain. Some regions and local authorities have been proactive (see pp158–161) on various environmental policy issues. Also, Spain has argued successfully for the possibility of using the EU cohesion and structural funds for environmental improvements, which has reduced the financial burden on the Spanish government for complying with EU environmental legislation.

However, in the case of climate change, despite the potential negative effects for the country itself, Spain has again been dragging its feet. When the EU discussed the target of stabilizing overall emissions of CO_2 at 1990 levels by the year 2000, Spain, together with the other cohesion countries, argued that it must be permitted to increase emissions rather than to pursue the stabilization objective. Spain insisted on the EU ratifying the FCCC as a bloc in order to allow it to do this, while still signing the convention. This stance was justified on the basis of Spain's lower per capita emissions and the below average level of economic development of Spain compared to other EU countries. Spain's stance is thus similar to that taken by developing countries in the global negotiations on emission reductions. The Spanish position takes for granted that the EU target is designed to balance out the expected growth in emissions from some of its Member States by the reductions in emissions from other Member States.

Initially, the Spanish government announced that it would limit emission increases to 25 per cent above the 1990 level. However, as it became clear that, because of a much lower than expected economic growth, emissions would grow at a slower rate, emission increases of only 13–15 per cent were considered more reasonable and now the government commitment is to a 10 per cent increase of CO_2 emissions between 1990 and 2000. However, it is unlikely that the Spanish government will agree to actual emission reductions post 2000 and it will expect to continue to 'free ride' within a general EU target, with reductions to come from the more developed EU Member States.

Spain ratified the FCCC on 21st December 1993, at the same time as the EU. The first Spanish communication to the Convention was prepared by the Secretary of State for Environment and Housing (SEMAV) of the Ministry for Public Works, Transport and the Environment (MOPTMA), with an outline of the different policies in place against climate change and a picture of Spanish emissions of greenhouse gases (MOPTMA, 1995). The SEMAV was also responsible for coordinating the National Climate Programme (PNC), which defined the application of different actions towards a better understanding of the climatic system and its relations with the different economic and social sectors.

The PNC was formulated by the National Climate Commission, an interministerial body with coordinatory functions and including representatives from the MOPTMA, the Ministry of Industry and Energy and the Ministry of Economy. The National Climate Commission must coordinate all actions concerning climate change in accordance with international criteria and must inform the government on the measures to be implemented. Most of these schemes and government bodies are, however, likely to experience significant changes with the recent creation of a Ministry for the

Environment by the new conservative government. In principle, the creation of a separate ministry should result in a greater emphasis being given to environmental issues because previously they were just one of many functions of MOPTMA.

The PNC places a particular emphasis on the potential effects of climate change on the country. According to FCCC guidelines the national reports should stress scientific and impact issues; they are also supposed to provide strategic proposals for government policies in the field. However, the current version of the PNC appears to describe the main lines of action without considering the real implementation issues. Effectively, climate change policy is seen to be largely related to Spain's environmental priorities, ie desertification, waste management, water management, biodiversity management and quality of the urban environment. It has not influenced energy and transport policy to any significant extent, and in these areas the programme mainly consists of measures which had already been agreed on for other reasons. None of the measures have been quantified in terms of their likely impact on CO$_2$ emissions, hence it will be difficult to evaluate their progress. The main actions with a CO$_2$ emission relevance are reviewed in the following sections.

ENERGY MEASURES

The Energy Policy Background

Energy policies have always constituted the backbone of Spanish industrial policies. A strategic area, it has been regulated strongly and directed by multiannual programmes, the National Energy Plans (PEN), of which the *PEN 91* is the most recent. The PENs set the basic framework for public and private actions in the sector, determining the direction and extent of public energy policies and their constraints. As indicated above, *PEN 91* is overoptimistic in its forecasts of the level of economic activity and is correspondingly inaccurate in forecasts of the main energy requirements. However, the Spanish energy domain can be characterized using the *PEN 91* together with the more realistic official updates.

In general, current Spanish energy policies do not consider environmental issues in a serious way. Environmental effects, particularly greenhouse gas emissions, seem to be subordinate to actions the priority of which is to research ways of reaching an energy capacity 'able' to facilitate the levels of economic development comparable with those of the wealthiest EU societies. Nevertheless, the *PEN 91* for the first time includes environmental protection among a number of different and sometimes contradictory objectives, namely:

■ cost minimization;
■ diversification;
■ use of national resources;
■ environmental protection (including CO$_2$ emission limitation).

PEN 91 presumes the continuation of the nuclear moratorium as it was implemented by the first socialist government in 1984 after years of strong public opposition to nuclear facilities.[2] The decision to keep or to cancel the moratorium is purely political, although the manifesto of the current ruling party states clearly that the moratorium will be maintained. Another constraint and an issue that may have influence on Spanish CO_2 emissions is the preferential system to encourage the use of Spanish coal, a policy designed to assure employment in large mining areas in the north of the country. The two mechanisms to foster consumption of Spanish coal currently in operation are minimum consumption quotas and subsidies to prices. These preferential systems are also considered by *PEN 91*.

The main aim of *PEN 91* is to reduce the use of primary energy per unit of GDP. It also considers introducing cleaner energy alternatives, such as renewables and natural gas, to enable diversification, efficiency, environmental protection and the use of national resources. To achieve the PEN objectives, the Savings and Energy Efficiency Plan (PAEE) has been developed by the Institute for Energy Diversification and Savings (IDAE) to attain these aims through four programmes:

■ Energy savings, with actions in industry, transport and services. It attempts to minimize consumption of fossil fuels and electricity.
■ Substituting oil and coal powered by natural gas powered facilities.
■ Cogeneration to foster joint production of heat and power. Cogeneration is assumed to optimize energy use industrial processes.
■ Renewable energy to encourage the introduction of alternative energy sources as a way of confronting climate change problems and contributing to long term security and diversification of energy supply.

Between 1991 and 1994, PAEE projects have achieved an approximate reduction of 4,500,000 tonnes of CO_2 emissions. The degree of compliance with the objectives for the year 2000 is high for cogeneration (80 per cent) and renewables (50 per cent), but much lower for savings and substitution (less than 20 per cent).

A CHP Boom

Within a very short period of time, Spain has become one of Europe's leaders in the development of CHP, which now meets almost 6 per cent of Spain's total electricity demand. This development has been government led, with the initial impetus provided by an energy conservation law of 1980, which lays down a legal framework that introduced tariff incentives. Later on, in 1986, information programmes, advisory services and third party financing arrangements provided by IDEA became important (COGEN, 1995). The PEN set a cogeneration target of 2222 MW_e for 2000, an increase of 76 per cent from the 1263 MW_e installed in 1990. The target had already been exceeded by the end of 1995, with 2656 MW_e installed. While

---

2 It stopped all future nuclear developments, except those in a very advanced stage.

most of the installed capacity is in the industrial sector, there is a growing application of cogeneration technologies to individual buildings (eg hospitals and hotels). The distribution of projects by industry shows that cogeneration is being used mainly by refineries, paper producers and textile industries. The distribution of projects by fuels shows that natural gas is involved in more than 60 per cent of the projects.

Some electricity utilities have become active in promoting CHP, in the form of investment and joint ventures, mainly because of the attractive rates of return offered. However, at the same time they have been opposed to the fact that CHP has introduced greater competition in the market. New legislation introduced in December 1995 has modified the tariff system for CHP surplus electricity, thus reducing the commercial attractiveness of CHP, in particular to larger plants. Buyback rates have been reduced and new definitions applied to the concept of surplus electricity. There appears to have been little consideration of the CO$_2$ benefits of CHP systems; its development remains driven by other considerations.

Neglect of End Use Energy Efficiency

Developments of end use efficiency are not very encouraging. The National Climate Programme does mention limited measures in this area, such as the application of existing laws on thermal insulation, but these have not been respected in the past. Some improvements in energy efficiency have occurred in industry in conjunction with the PAEE. However, as already mentioned on page 151, energy efficiency has not been a priority in Spain, resulting in an upwards trend in energy intensity. The public sector budget for improving end use efficiency has seen severe cut backs in recent years. Spending fell by 43.7 per cent between 1992 and 1993 and a further 33.8 per cent between 1993 and 1994. Most cut backs have affected the programmes for the commercial and industrial sectors. At the same time, electricity utilities have shown little interest in this area.

There are some signs of improvement. The Ministry of Energy is encouraging electricity producers to carry out DSM projects and has provided 5000 million pesetas for this thus far. Furthermore, in April 1995 the government decided to increase the funding for the PAEE energy efficiency measures. For the period 1995–2000, 106 billion pesetas are to be available (IEA, 1996). The programme will focus on buildings and the transport sector.

To some extent there has been a natural incentive for energy efficiency investments. Spanish electricity prices have been heavily regulated traditionally and in general are higher than in other European countries. The Spanish economic expansion of the 1980s and the monopolistic conformation of the electricity system induced a significant gap between domestic and international electricity prices. This may have encouraged efficient patterns of electricity demand and use (although the evidence here is limited considering the worsening figures for energy intensity), but price divergences could also have had a negative effect on industrial competitiveness and may render difficult the introduction of parallel climate change policies. Top of the agenda of the new government is to deregulate the electricity generating industry, with the aim of lowering energy prices through competition. Hence, any previous incentive for energy efficiency is likely to be reduced.

A Positive Outlook for Renewables

Spain has a considerable potential for renewable energy sources and limited fossil fuel reserves, factors which are in principle beneficial to the development of renewable energies. The climate conditions make wind and solar sources feasible for energy production. While the highest wind speeds experienced on mainland Spain are restricted to 2.6 per cent of the land area, over three quarters of the land area experiences favourable wind speeds in the band 5–6 m/s (European Commission, 1994). There is also unexploited potential for hydropower, although the remaining sites for large head hydropower would be environmentally difficult and costly to develop. However, there is potential for small head hydropower plants. Furthermore, biomass and municipal solid waste fired capacity could be developed. Solar collectors and PVs are obviously favoured by the climatic conditions but have suffered from high costs.

Currently, renewables account for 5.17 per cent or 4.94 Mtoe of Spain's 95.54 Mtoe primary energy consumption. Similar to most of the nations reviewed in this volume, hydropower is the largest renewable energy source, accounting for 2.4 Mtoe or 49 per cent of the total renewables contribution of 4.94 Mtoe, closely followed by biomass with 2.39 Mtoe (48.4 per cent). In 1991 the PAEE called for investments in renewables which would lead to a yearly addition of 1.1 Mtoe by the year 2000. These investments would be spread across a whole series of different renewables, some of which are already close to (or even surpass) their initial year 2000 targets while others are far from achieving it. As in a number of other EU countries, wind power is turning out to be the most attractive renewable resource. IDAE has estimated a potential capacity of 2400 MW, with the greatest potential in the region of Galicia (MINER, 1994). Wind farms have surpassed their set targets of 38,200 toe/year to 110.3 per cent and so have PVs (437,100 toe/year – 112.4 per cent). Geothermal, on the other hand, has only reached 4.4 per cent of its target of 443 toe/year.

Increasingly, renewables are seen to play a large role in making many of Spain's islands more sustainable. The best example of this is the Canary Islands' autonomous administration which has put forward a six year plan (1996–2002) to increase the contribution of renewables seven fold from the current level of 15,486 toe/year to 121,000 toe/year. This goal would represent 3.6 per cent of the region's primary energy consumption.[3] The possibility of using of the EU structural funds (Narbona, 1994) is becoming important for the development of renewable energies in Spain, especially at the regional level.

---

3 *Renewable Energy Report*, No 9, 17th February 1996, p 10.

TRANSPORT POLICIES

Almost one quarter of Spanish CO$_2$ emissions were caused by the transport sector in 1993, and most greenhouse effect precursors such as NO$_x$, CO or volatile organic compounds (VOCs) were emitted by these sources. This was mainly due to the considerable increase in mobility observed in Spain during the past 20 years, a trend found in most European countries. All predictions forecast a substantial increase of activity in this sector, with a simultaneous growth of its emissions.

Between 1972 and 1990, road transits more than trebled and air transits more than quadrupled, while railway transit remained stable. Figure 9.3 illustrates the fast growth in road based transport. Transport of commodities by road, road transport by coach and the use of suburban trains underwent a sustained growth. Other transport sectors, especially interurban railway, went through processes of stabilization or even decrease. Spanish public transport compares poorly with other European countries, with a railway system that shows severe signs of underinvestment.

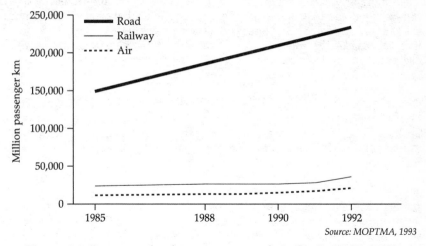

Source: MOPTMA, 1993

Figure 9.3: Passenger km by transport mode in Spain (1985–1992)

Transport policy over the past decade or so has been essentially road focused. In 1983, a major construction programme was launched under the General Plan for Roads, including the building of 5000 km of motorway. It is likely that continued Spanish economic integration into the EU, together with the predictable processes of economic growth, will cause a large increase in mobility. Existing trends and official actions indicate that almost all the projected increase in transit will be absorbed by road transport. This phenomenon may be strengthened by the general trend towards out-of-town shopping centres and office developments. This is already happening in places like Madrid, although its extension to other regions seems to be less remarkable than in other European countries.

Despite the worrying trends regarding the growth of CO_2 emissions from the transport sector, there are no meaningful policies aimed at abating greenhouse emissions from this sector or at climate change strategies in this field. Most government actions encourage rather than restrict greenhouse gas emissions from the transport sector (for instance, through public investments in roads or low fuel tax levels). In fact, the new government's manifesto is especially keen on the building of new motorways through a new Programme of Infrastructures.

There are, however, some policies aimed at improving public transport, which should lead to some reductions in CO_2 emissions from road traffic. For example, the Transport Plan for Large Cities (1990–1993) encouraged the modernization and improvement of public transport. There were investments in suburban railway infrastructures in Madrid, Barcelona, Seville, Valencia, Oviedo and Gijón. These railway developments included out-of-town parking spaces to reduce the number of cars entering the cities.

Other government investments include the metro systems in Bilbao and Seville. Madrid's metro is also being enlarged. There has also been a tax exemption for fuel oil used by railways since 1993 which aims to encourage rail transport and to promote the substitution of electricity by fuel oil within the railway system. This is less damaging from a climate change perspective because of the importance of coal fired power stations within the electricity generating industry.

REGIONAL AND LOCAL AUTHORITIES AND THE PRIVATE SECTOR

Spain is a highly decentralized country with 17 regions and more than 2000 municipalities. Environmental policy responsibilities are thus shared by a number of tiers of government:

■ European Union;
■ National government;
■ Regions, or *'comunidades autónomas'*;
■ Municipalities.

The Spanish Constitution states that:

■ The national government has full responsibility for the definition of basic environmental legislation, with a coordinatory function over the regions.
■ Regions have full powers in the development of basic environmental legislation, management of environmental programmes and inspecting and penalizing activities.
■ Municipalities have some local environmental responsibilities, usually coordinated and funded by the regional authorities.

'Comunidades Autónomas'

As at the national level, there are no specific climate change policies at the regional level. Unlike the national government, regional authorities do not even see the issue as relevant. Policies in other spheres may influence greenhouse gas emissions but in most cases they do not consider climate change problems at all. In addition, *comunidades autónomas* have little scope to legislate with respect to climate change. They have very few responsibilities in the energy sector as this is considered to be a national matter.

There have been a number of meetings on climate change between representatives from the national government (MOPTMA) and officials with environmental responsibilities from the regions. These meetings have been held by the coordinating body for national and regional authorities, the 'Regional Committee on Environment'. Regional representatives noted that these meetings were only designed to provide information on current national strategies and to comply with international commitments.

For this review, the author interviewed representatives from the Catalonia and Galicia regions. Catalonia has an environmental ministry, while Galicia has a less organized environmental agency and a poor environmental record. In both cases the regional officials indicated that their priorities were in 'traditional' environmental areas, especially the management of solid and liquid wastes. The Catalan representative indicated a reluctance to develop independent environmental policies. The region has, however, implemented strong measures to boost public transport in Barcelona, actions which could reduce CO_2 emissions. Catalan building standards are stricter than those at the national level, improving energy conservation. The regional government has also encouraged urban planning, enlarging the underground transport system in Barcelona, and recently introduced high speed rail links to the French network. Finally, the Catalan government also attempts to facilitate environmental auditing and voluntary agreements with the private sector.

The Galician government representative felt that decisions on this global issue should be taken by Spanish or European authorities, but that policies must be developed locally. Several large coal fired power plants are located in Galicia and the regional government has recently introduced an environmental tax on SO_2 and NO_x emissions. The tax, strongly opposed by the Spanish electricity industry, could have positive effects for CO_2 emissions but its very low rate suggests it is designed principally as a revenue raising instrument.

Transport policies in other regions along the lines of those adopted in Catalonia could foster a reduction of CO_2 emissions. However, efforts to develop road networks have had predictably negative impacts. Regional policies on energy efficiency and alternative energy sources have had more effect on overall CO_2 emission reductions. There are several regional organizations that deal with renewable energies and energy savings, for instance the Catalan Institute for Energy (ICE). The ICE works mainly in the industrial and transport sectors and has contributed to a significant expansion of cogeneration, biomass use, and small head hydropower capabilities in the Catalan industrial sector. It has also obtained good results with PV technology in rural electrification.

Local Authorities

Municipal governments are responsible for enforcing the basic legislation on atmospheric pollution. Local authorities are allowed to change emission limits as long as they are within the national legislation. In order to obtain information on the current state of climate change policies and future directions at the local level, the author visited the Barcelona, Madrid and Vigo municipalities. Barcelona and Madrid are the largest Spanish cities, each with a population of around 3,000,000. Vigo is the largest city in Galicia, with approximately 250,000 inhabitants. A representative from the environment department was interviewed in each place.

Vigo appears to be the least interested in climate change policies. The municipal officials indicated that this is an issue for authorities at higher levels. They did say that there is room for subsidiarity in policies against climate change as long as extra funding is made available by regional, national or European authorities. There is no climate change policy or strategy in Vigo, and even the effects from collateral policies are quite circumscribed.

Madrid has an intermediate attitude towards climate change – it attended the Berlin summit in April 1995 and intends to sign ICLEI's 'Heidelberg Declaration', committing itself to a 20 per cent reduction with respect to 1987 CO_2 emissions for 2005. Although Madrid has not developed a climate change strategy so far, there are some policies aimed at reducing CO_2 emissions and several municipal policies could have a contributory effect. Among the former could be included some basic lines of the General Urban Plan (PGOU), reconciling all urban enlargements with vigorous public transport schemes and setting tax exemptions for new buildings with significant energy savings and/or use of renewable energies. Policies with collateral effects comprise expansion of green areas (CO_2 sinks), encouragement of public transport (lanes for vehicles with more passengers, intermodal transport exchange centres and expansion of the underground), use of waste for electricity generation and municipal grants for the substitution of polluting and inefficient heating systems.

Finally Barcelona constitutes an interesting case, with a remarkable development of climate change policies and strategies. Barcelona attended the Rio (1992) and Berlin (1995) summits and signed ICLEI's 'Heidelberg Declaration' in September 1994. It is also the only Spanish city which is currently calculating an inventory of CO_2 emissions. Barcelona's head of the environmental programme indicated that subsidiarity should be fully applied to climate change policies. He considered that cities constitute a very important factor in attaining effective climate change strategies and demanded more legal responsibilities (eg through a Spanish Act for large cities) and more resources to execute those strategies. Policies in place include the introduction of significant savings in public lighting and the reduction of energy consumption from municipal surfaces and buildings (Ajuntament de Barcelona, 1994). The municipality has obtained these positive results after a general implementation of energy audits, and uses these savings to promote more energy efficiency. Actions in the transport sector include an ambitious programme of intermodal exchange centres, a progressive introduction of natural gas buses and several municipal regulations to minimize transport requirements by promoting housing and economic activities in all areas of the city.

The Private Sector

In a setting where the lack of industrial competitiveness is perceived to be the cause of the Spanish malady, vigorous competition from other EU Member States and EU-induced environmental regulations appear to worsen the situation. With this background the attitude of the Spanish industrial sector towards environmental issues, and particularly towards climate change, can be summarized as follows:

■ Spanish firms are reluctant to accept demanding environmental regulations because they see them as a threat to their competitiveness. They will not adopt voluntary environmental policies unless there are no conceiveable possibilities of a financial or technical burden.
■ When obliged to implement some environmental measures, Spanish industries expect support from the public sector.
■ The Spanish industrial sector is especially opposed to any modification of energy prices for environmental reasons. These price changes are seen as very damaging to competitiveness, especially taking into account that Spanish energy prices are thought to be higher compared to those in other competitor countries.
■ The industrial sector prefers to seek solutions to immediate pollution problems rather than to embark on climate change policies, considered to be a luxury for a country with comparatively low levels of economic activity and emissions. Free riding by other competitors is always in the mind of the Spanish industrialists.

CONCLUSIONS: ASSESSMENT OF CURRENT STRATEGIES AND FEASIBILITY OF FURTHER ACTION

Spain's exemption from the FCCC stabilization commitments through the informal EU target sharing arrangement informs and determines most Spanish actions in the area. In this chapter it is observed that Spain lacks a general climate change strategy, although some unconnected and partial measures do exist. This absence of a climate change policy is particularly noticeable in the area of transport.

Most Spanish policies with effects on climate change abatement have originated from the national government, and show poor coordination with regional and local bodies. These actions have been mainly implemented in the industrial and residential areas, with a strong public funding component. Voluntary agreements have also been a preferred option for both the regulated and the regulators. Economic instruments for climate change policies (markets and/or taxes) have not been considered at all.

At the present time there do not seem to be many options to implement climate change policies in Spain. In order to attain reductions in greenhouse gas emissions, both regulators and regulated consistently prefer voluntary agreements to regulatory interventions. The origin of this behaviour appears to be in the low (comparatively) levels of Spanish CO_2 emissions, together with the severe problems faced by many sectors of the Spanish economy. Consequently, the adoption of compulsory and ambitious measures to cope with greenhouse gas emissions looks unlikely. It is likely that future developments will follow similar lines to those observed to date, principally:

■ Climate change commitments generally will be induced exogenously. Most official actions will aim at obtaining emission abatements from residential and industrial sources, while transport policies will be scarcely affected. Therefore, public intervention and instrument use will be limited in their extent and scope of application.

■ Spanish authorities will favour consent-seeking attitudes from industry, to avoid sudden structural changes at even moderate costs to the Spanish economy.

■ Climate change policies will rely on the copious Spanish and EU public grants to the private sector, allowing for easier transition processes. There might be some incompatibilities between these funding policies and the 'Polluter Pays Principle', although most actions will stress their R&D nature, with energy related rather than environmental objectives.

■ Energy taxes or carbon permits will not be a prominent policy option. Their distributional and economic effects are likely to be high, and their environmental effectiveness may be reduced by the special features of the Spanish energy sector. There might be tax increases for automotive fuels, although these will mainly have revenue raising purposes and will not offset the negative effects that arise from other public policies in the transport field.

In the following concluding sections, we summarize the factors that may encourage a reduction or stabilization of emissions, as well as those which are likely to act as obstructive factors. The comparative strength of these factors and the level of future economic growth will determine the evolution of Spanish CO_2 emissions.

Enabling Factors

■ There is room for clear improvements in transport and energy related emissions through:
 – modification of public policies in the field;
 – introduction of new market or institutional frameworks;
 – introduction of precise instruments aimed at energy efficiency and climate change abatements.

■ Energy policies have been directed mostly at supply, with positive effects arising from the extension of renewable sources and natural gas. By acting on the demand segment, there may be extra gains.

■ There are good conditions for the use of renewable energies in Spain.
■ There is a growing consideration and application of win–win strategies by the private sector.
■ Policy developments for the residential field and urban planning actions are likely to be significant.
■ There is increasing social awareness on the risks of human-induced climate change phenomena.

Obstructing Factors

■ There is little official interest in the design and implementation of environmental policies.
■ Coordination within and between levels of government seems to be fairly ineffective in regard to climate change policies.
■ There is an increasing demand of transport and energy from the development processes of the Spanish economy. Transport figures seem to be particularly out of control.
■ There are imperfect markets for transport and energy in Spain. This situation may induce costly results from the application of climate change policies.
■ Spanish energy policies have to adjust to sociopolitical components, with the compulsory use of poor quality Spanish coals.
■ Spanish energy markets display large demand rigidities, with little incidence of price policies.
■ The nuclear moratorium and technological limits of renewable energies are likely to foster CO$_2$ emissions.
■ Industrial sectors have shown poor environmental performances.

It has already been noted that Spanish society is increasingly concerned with global warming phenomena. Contemporary climate disfunctions are thought to be the first indications of acute climate changes that would severely affect Spain. The Spanish environmental movement has played a significant part in publicizing these concerns, with active policies to encourage social awareness. In fact, climate change constitutes a top priority for Spanish environmentalism. Although the status of the Spanish environmental movement is not comparable to that observed in other European countries, it has a growing influence and audience within Spanish society. It thus appears that Spanish authorities will face a growing pressure from the voting public and from the environmental movement. In the next few years this combined action may encourage a more active stance on the part of successive Spanish administrations.

Concerning future developments, it is likely that there will be few changes to current trends. Policies aimed at climate change abatement will usually follow EU requirements (regulations and directives), although the Spanish government will attempt to circumscribe most actions to areas outside transport and energy supply. Therefore, greenhouse gas emissions will mainly depend on the evolution of the Spanish economy.

Overall, there appear to be few reasons for optimism for a stabilization of greenhouse emissions in Spain. Lax international commitments and internal realities will encourage climate change inducement rather than abatement. Although we have shown that there are enabling factors that favour the introduction of more determined policies, the obstructing factors are likely to impede any ambitious move. It is imperative that any sensible climate change strategy should include the removal of environment unfriendly policies, especially in the transport field, with a simultaneous introduction of institutional changes in energy markets. Economic instruments directed at reducing greenhouse gas emissions should be introduced at a second stage.

REFERENCES

Ajuntament de Barcelona (1994) *Programas de Actuación para una Política Medioambiental en Barcelona*, Barcelona, unpublished

Aquilar, S (1993) 'Corporatist and statist designs in environmental policy: the contrasting roles of Germany and Spain in the EC scenario', *Environmental Politics*, Vol 2, no 2, pp223–47

COGEN (1995) *The Barriers to Combined Heat and Power in Europe*, COGEN Europe, Brussels

European Commission (1994) *The Renewable Energy Study, Annex 2, Country Profiles*, DG XII, Brussels

IEA (1996) *Energy Policies of IEA Countries. Spain 1996 review*, OECD/IEA, Paris

MINER (1994) *Las Energías Renovables en España. Balance y Perspectivas hacia el 2000*, Secretary General of Energy and Mining Resources, Madrid

MINER (1995) *Plan Energético Nacional 1991–2000: Balance 1995 y Perspectivas 2000*, Ministry of Industry and Energy, Madrid.

Ministry of Industry (1991) *Plan Energético Nacional 1991–2000*, Secretary General of Energy and Mining Resources, Madrid

MOPTMA (1993) *Los Transportes y las Comunicaciones. Anuario 1992*, Institute for the Studies of Transport and Communications, Madrid

MOPTMA (1994) *Informe de España a la Convención Marco de las Naciones Unidas sobre el Cambio Climático*, Secretary of State for Environment and Housing, Madrid

MOPTMA (1995) *Programa Nacional sobre el Clima*, Secretary of State for Environment and Housing, Madrid

Narbona, C (1994) 'Medio ambiente. Aplicación a la política medioambiental del instrumento financiero de cohesión', *Presupuesto y Gasto Público*, No 14, pp239–257

OECD (1993) *OECD Environmental Data. Compendium 1993*, OECD, Paris

Partido Popular (1996) *Con la Nueva Mayoría*, Manifesto, Madrid

Chapter 10 | SWEDEN: THE DILEMMA OF A PROPOSED NUCLEAR PHASE OUT

Ragnar E Löfstedt

INTRODUCTION

In the area of climate change policy Sweden can be characterized as a follower. That is to say it has agreed to stabilize CO_2 emissions by the year 2000 as stipulated in the FCCC, but it will not set an example and cut CO_2 emissions substantially post 2000. Although Sweden, just as France (see Chapter 8), had reduced CO_2 emissions substantially prior to the climate change debate phase (it reduced emissions by 40 per cent between 1970 and 1990) the nation is finding it increasingly difficult to make further CO_2 reductions. It has put forward a CO_2 tax which is among the highest in the world, yet since 1990 CO_2 emissions have actually increased. The whole climate change policy problematic has also been worsened by the uncertainty as to whether Sweden's 12 nuclear reactors will be phased out and, if they are, as to what types of energy sources will replace them. In Sweden strategies to combat climate change, particularly CO_2 emission reduction policies, have been developed almost exclusively as a product of energy policy considerations. Therefore this chapter pays particular attention to the country's energy sector.

SWEDEN'S GREENHOUSE GAS EMISSIONS

By far the most important greenhouse gas in Sweden is CO_2. Its emissions are set to increase by almost 8 million tonnes on 1990 levels by the year 2000, even with the CO_2 abatement measures currently in place. In calculating the GWP[1] of Sweden's greenhouse gas emissions, CO_2 accounts for 55.7 million tonnes of the 76 million tonnes total in 1990 (or approximately 73 per cent), while CH_4, the second most important gas, contributes only 8.1 million tonnes (Table 10.1). Given the projected increases in CO_2 emissions, mainly in the traffic sector, and the planned decreases or stabilization of most other gases, the relative importance of CO_2 is set to increase further.

Table 10.1. *Summary of Sweden's Total Greenhouse Emissions*

| Emission | Million tons of CO_2 equivalent | | | |
	1990	1994	2000	2005
CO_2*	55.7	58.4	63.8	67.9
CH_4	8.1	6.4	7.4	7.2
N_2O	5.0	4.3	4.3	4.0
HFC	0.0	0.1	2.6	2.6
Fluorocarbons	0.4	0.1	0.4	0.4
Sulphurhexafluoride	1.0	0.5	1.0	1.0
Total	70.2	69.8	78.7	82.1

* CO_2 emissions as calculated by NUTEK, the Swedish energy efficiency agency.

Source: Swedish EPA, 1995

Currently half the N_2O emissions originate in the agricultural sector, coming largely from the use of organic and inorganic nitrogen fertilizers. Reductions in emissions are planned for the period up to 2005 through changes in agricultural practices, including reducing the number of hectares farmed and the number of grazing animals per hectare, and restricting the practice of slurry spreading (Swedish Ministry of the Environment and Natural Resources, 1994a).

The Swedish goal is to reduce CH_4 emissions from landfill sites by 30 per cent by the year 2000, contributing to a 2 per cent net decrease in greenhouse gases based on 1990 levels. This goal appears likely to be met as, by 1995, there had already been a reduction in CH_4 emissions by 20 per cent based on 1990 levels [Swedish Environmental Protection Agency (Swedish EPA), 1995]. CFCs are to be completely phased out by 1995 and replaced by HFCs. However, HFCs are potent greenhouse gases and it is envisaged that they will represent 3.2 per cent of Sweden's GWP by the year 2005. Finally, there is sulphurhexafluoride (SF_6), which is primarily used for insulating gas and electrical equipment. Currently there are no substitutes available for SF_6 so consumption will continue at current levels (Swedish Ministry of the Environment and Natural Resources, 1994a; Swedish EPA, 1995). Overall, in this scenario, only CO_2 and HFC emission levels are set to increase up to 2005.

1 GWP is calculated using the formula put forward by the IPCC. All figures are converted into million tonnes of CO_2 equivalent.

CO_2 Emissions

Between 1970 and 1990 Sweden reduced its CO_2 emissions by 40 per cent (Table 10.2). The main reason for this was the reduction of oil usage through energy conservation measures, electricity generated from nuclear power and biomass.

Table 10.2. *Emissions of CO_2 by Sector (1970–1990)*

Year	Electricity and heat production (Mt)	Traffic (Mt)	Industry (Mt)	Total (Mt)
1970	80.0	16.0	4.0	100.0
1975	65.0	17.0	3.0	85.0
1980	57.3	21.9	2.8	82.0
1985	41.7	22.3	2.9	66.9
1990	32.0	23.0	5.0	61.2*
1990**	30.0	21.4	3.6	55.7*
1993	30.23	26.5	4.0	60.73
1994	31.57	27.79	4.14	63.5
1994**	31.3	22.4	4.1	58.4*

*Also includes CO_2 emissions from agriculture, refuse and various sources not accounted for in the other categories.

** CO_2 emissions as calculated by NUTEK and reported in Swedish EPA (1995).[2]

Source: 1970–1985 data based on information from the Swedish EPA and the Swedish Bureau of Statistics and cited in Swedish Parliament (1992–1993); 1990 data based on Swedish Ministry of Environment and Natural Resources (1994a); 1993 and 1994 data based on the Swedish Bureau of Statistics (1994, 1995).

Owing to the possible phase out of nuclear power as stipulated in a 1980 referendum on the future of the industry, there is a general unwillingness to predict CO_2 emissions beyond the year 2005. However, based on the premise that nuclear power is still in place up to 2010 the tentative predictions in Table 10.3 have been made.

Of the sources of emissions analysed in recent years, the energy and traffic sectors have probably received the most attention. These are discussed in depth on pp168–178. The greatest uncertainty in the government's predictions is the future of nuclear power, which currently generates 50 per cent of Sweden's electricity requirements. According to a 1980 parliamentary decision, all of Sweden's nuclear power plants should be phased out by 2010. If this were to be the case CO_2 emissions would increase

2 In Sweden the CO_2 emissions data and overall greenhouse gas emissions are continuously being refined. There are several reasons for this. The calculations on CO_2 emissions made by NUTEK, first put in place in the Swedish EPA 1995 (October report), give different numbers for CO_2 emissions than the data from the Swedish Central Bureau of Statistics. In order to avoid confusion the CO_2 numbers from both sources are included for 1990 and 1994.

Table 10.3. *CO_2 Emissions in the Years 1990–2010 (NUTEK's Calculations)*

Source	1990	1994	2000	2005	2010*
Energy generation sector	6.9	8.7	10.9	13.7	16.5
Transportation	21.4	22.4	25.3	26.7	28.1
Industry	13.7	14.2	13.1	13.7	14.3
Residential and commercial	9.4	8.4	8.4	7.7	7.0
Industrial processes	3.6	4.1	5.0	5.0	5.0
Other sources	1.2	1.2	1.1	1.1	1.1
Totals	55.7	58.4	63.8	67.9	72.0

*2010 numbers extrapolated from the 2005 predictions.

Source: Swedish Ministry of the Environment and Natural Resources (1994a) and Swedish EPA (1995).

dramatically up to 2015. Some studies indicate CO_2 emission increases of over 160 per cent in a high economic growth scenario or 70 per cent if economic growth is low (Swedish Energy Administration, 1989). Other studies suggest that if all nuclear electricity was replaced by natural gas, CO_2 emissions would increase by 50 per cent based on 1990 levels. If the nuclear capacity was to be replaced by coal generated electricity, emissions would increase by 100 per cent, to more than 120 million tonnes CO_2 annually (SNCC and Swedish EPA, 1995).

POLICY DEVELOPMENTS IN THE AREA OF CLIMATE CHANGE

Background to Sweden's Climate Change Strategies

Sweden's climate change policy was initiated in 1988 as a part of the debate between the pro and anti nuclear groups. This debate has its roots in the controversial decision to phase out nuclear power, which represents 50 per cent of Sweden's electricity production (65 TWh), by the year 2010 following a referendum in 1980. Since 1980 there has been debate as to whether a phase out is economically possible by the year 2010. Following a political compromise reached in 1991, where it was decided to phase out nuclear power only when it was economically and socially possible, the political debate developed into a fully fledged climate policy with the highest CO_2 taxes in the OECD.

Sweden's climate change policy, similar to most other nations, is based on a wait and see approach. Policy makers had been following the work on climate change, especially that relating to the 1985 Villach Conference (WMO, 1986) and the 1987 Brundtland Report (*Our Common Future*) (World Commission on Environment and Development, 1987), but until 1987 the Government was reluctant to make any commitments to halt climate change based on this research (Swedish Parliament, 1987–1988).

The first policy to stabilize CO_2 emissions originated from a 1988 proposal by the Swedish Parliamentary Select Committee for Agriculture (*Jordbruksutskott*), which was demonstrably concerned about the impacts of climate change (Jordbruksutskottet, 1987–1988; Klefbom, 1993). Following a series of political debates concerning the importance of phasing out nuclear power or reducing CO_2 emissions, parliament agreed to stabilize CO_2 emissions based on 1988 levels. However, the decision was by no means unanimous, and as with previous parliamentary disagreements on energy policy, the government decided to set up an energy commission to examine the question. The recommendations of the Commission were integrated into the 1991 Energy Policy Act (Swedish Parliament, 1990–1991). This Act stated that the nuclear phase out would not begin in 1995–1996 and nuclear power could possibly remain after 2010. It would be phased out when the Swedish economy and/or workforce would not be adversely affected. Most importantly for climate change policy, the Act also opened the way to remove the 1988 CO_2 emissions ceiling as it was argued that it was neither economically nor technically feasible to maintain the ceiling at 1988 levels. The Commission believed that these had already been surpassed by 1990, despite statistics to the contrary from the Swedish Environmental Protection Board (Eriksson, 1991) and decided that Sweden should adhere to EU and EFTA (European Free Trade Association) guidelines calling for a stabilization of CO_2 emissions based on 1990 levels by the year 2000 (Swedish Parliament, 1990–1991).

Probably the most important historical development for Sweden's climate change policy was the 1992 UNCED. The Swedish Government was an enthusiastic proponent of UNCED and signed and ratified the Climate Change Convention finalized at the meeting that called for a stabilization of CO_2 emissions based on 1990 levels by the year 2000. As a way of meeting the commitment to stabilize CO_2 emissions the Conservative-led government introduced a climate change bill in 1993 which called for a reduction in CH_4 emissions and investment in renewable energy sources and energy conservation in Sweden, and the development of a Joint Implementation pilot programme with the Baltic States.

The Main Actors in the Area of Climate Change

Sweden's climate change policy has been driven by the Ministry of Environment and Natural Resources and its research arm, the Swedish EPA. These two bodies are made up of politicians, experts and civil servants. Currently, the Environmental Minister, Anna Lindh, and the Vice Environmental Minister, Dr Måns Lönnroth, are the main policy makers. As all sectors using fossil fuels contribute to climate change, other Ministries play a role in forming Sweden's climate policies. These include the Ministry of Transport and more importantly the Ministry of Industry and Commerce and its research arm NUTEK (Swedish National Board for Industrial and Technical Development). This latter Ministry is currently playing a particular important role, re-examining the feasibility of phasing out nuclear power and how this will affect CO_2 emissions. The opposition parties have also had an important role. It was the Conservative, Liberal and Centre parties who put global warming on the agenda in 1988 in the first place.

Finally, there are the environmental NGOs and the public. Environmental NGOs, of which Greenpeace and the Swedish Society for the Conservation of Nature are dominant, have played only a minor role in influencing the policy making process. As with the nuclear power debate in the 1970s, they have been unable to strongly influence the climate change debate. Rather, they became involved with climate policy once the opposition parties identified it as a key issue in the late 1980s, and even then they were divided. As recently as 1990 one of the directors of the Society for the Conservation of Nature argued that it should concentrate on pressing for the phase out of nuclear power rather than on stabilizing CO_2 emissions (Kågeson, 1990a, 1990b). Additionally, environmental NGOs have had credibility problems related to climate change as their views on it are seen to be rather radical; they are currently arguing that nuclear power can be phased out by 2010 and at the same time CO_2 emissions can be reduced 20 per cent by the year 2005. Most policy makers believe that such a strategy would destroy Sweden's economy.

The public have played virtually no role in influencing the climate change policy process. The main reason for this is that the Swedish public, like the American (Kempton, 1991) and Austrian publics (Löfstedt, 1993c), know little about the causes, effects and outcomes of climate change, often confusing it with the ozone hole (Löfstedt, 1991, 1992, 1993a, 1993b). This is not to say that they are unaware of it. Surveys have shown that over 90 per cent of those questioned have heard of climate change, but that they do not know what it is (see Löfstedt, 1991, 1992). This suggests that the Swedish public needs information about global warming to be supplied in a more systematic way.

CURRENT CLIMATE POLICY – HOW TO REDUCE CO_2 EMISSIONS

Sweden's climate policy is based on a series of differentiated policy measures. At the national level the most important include the CO_2 tax, investments in renewables and energy conservation (Table 10.4).

The CO_2 Tax

As discussed on pp168–169, in 1988 the Social Democrats raised the possibility of levying a CO_2 tax to reduce emissions. This became law in 1990, when the then Energy and Environmental Minister Birgitta Dahl argued in Parliament that, due to the threat of climate change, a tax on fossil fuel use of 250 SEK per tonne CO_2 would be sensible (Swedish Parliament, 1989–1990). The tax was placed on all fuel use except that used in power generation and peat at a rate of 250 SEK per tonne of CO_2 emitted. At the same time the government reduced the other energy taxes by 50 per cent. Energy intensive industry, however, was able to obtain a tax concession, whereby an energy and carbon tax ceiling was set at 1.7 per cent of the sales

Table 10.4. *Measures to Reduce CO_2 Emissions under Sweden's Climate Policy*

Measure	Reduction to be achieved (Mt)
CO_2 tax (energy sector)	5.3
Gasoline tax and CO_2 tax in the transport sector	2.2
Energy conservation programme	2.1
Investments in biofuels	0.6
Other	0.2
Total	10.4

Source: Swedish Ministry of the Environment and Natural Resources, 1994a[3]

value of production (Swedish Ministry of the Environment and Natural Resources, 1994a).

In 1993 the energy tax system changed again. The Conservative-led government, concerned about the continued recession and the effect that this was having on Swedish industry, abolished the general energy tax completely for industry and lowered the CO_2 tax to 25 per cent of the amount levied on other sectors. This, it was argued, reduced Sweden's economic disadvantage compared to other industrialized countries that had much lower, or even non existent, energy taxes.

As a result, in 1994 the CO_2 tax was 83.2 SEK per tonne CO_2 for industry and 322.8 SEK per tonne for all other users. So that the tax is not undermined by inflation, it has been index adjusted at 4 per cent a year for the period 1994–1998. In March 1996 the CO_2 tax was changed once again as the Social Democratic government proposed an increase of the CO_2 tax for industry to 50 per cent of other users, since the Environmental Minister felt that such a change was now feasible as the industry was no longer in recession. As a result of this tax increase, industry today pays 135 SEK per tonne CO_2 and other users pay 370 SEK (Table 10.5).

Table 10.5. *Swedish CO_2 Taxes*

Year	Industry (SEK per tonne of CO_2)	Other users (SEK per tonne of CO_2)
1991	250.00	250.00
1994	83.20	322.80
1996	135.00	370.00

3 The Swedish government believes that these measures will reduce CO_2 emissions by 10.4 million tonnes by the year 2000. If these measures were not introduced Sweden would emit 74 million tonnes CO_2 by the year 2000.

The initial change in the taxation structure shifted the burden of CO_2 emission reduction away from the industrial sector, thereby in effect encouraging industry to use more fossil fuels and resulting in an increase in CO_2 emissions. Independent sources show that between 1992–1994 CO_2 emissions grew by 14 per cent in the industrial sector mainly due to a 20 per cent increase in oil use (Bjerström, 1994). However, this is likely to be reversed following the government's decisions to double CO_2 tax on industry.

In short, taxation has become Sweden's main mechanism for controlling CO_2 emissions. In the climate change strategy it is projected to deliver more than 50 per cent of the CO_2 reductions (see Table 10.4, p297), making its contribution more important than the investment programmes in biomass, wind power or energy conservation discussed below. However, it is unclear how successful this tax will be in reducing emissions. When the tax was introduced in 1991 other energy taxes were reduced by 50 per cent. Studies indicate that since the CO_2 tax was introduced the tax burden for industry has actually decreased and that CO_2 emissions have increased (Bjerström, 1994). This may now have reversed following the recent CO_2 tax increase in the industrial sector, but at the time of writing (June 1996) it is too early to tell if this is indeed the case.

Furthermore, although the CO_2 tax has led to a greater use of biomass in district heating, this has been partially offset by an increase in fossil fuel use in the forest industry. The CO_2 tax makes it more economic for forest companies to sell pine oil to district heating plants, and use fossil fuels for energy. As a result the CO_2 tax in this case has not led to reduced CO_2 emissions, but rather to a change in where the CO_2 is emitted from (Swedish Ministry for the Environment and Natural Resources, 1994b).

This is not to say that the CO_2 tax has been completely ineffective. The consumption of gasoline and diesel has decreased significantly due to the higher prices (NUTEK, 1994a), and there has been a definite shift in district heating to biomass from other energy sources. Based on recent predictions, which point out that this switching is cost effective, these trends should continue at least in the short-term (Agricultural Ministry, 1993). The question is how long the CO_2 tax can influence gasoline consumption and the types of fuel used in district heating. Current trends indicate that gasoline and diesel consumption will continue to increase in the medium term (SNCC and Swedish EPA, 1995); based on current consumption levels there is only a limited amount of district heating capacity that can be switched from coal and oil to biomass (13.6 TWh according to the Swedish Heating Association, 1994). For CO_2 taxes to have a significant impact on the level of emissions the tax has to be increased and expanded to cover all sectors: manufacturing and energy intensive industries, for instance, should not be exempt from a full CO_2 tax.

Environmental groups and some political parties, including the Christian Democrats, favour tax switching mechanisms where taxes on labour are switched to energy consumption or other environmentally harmful activities. Advocates of this approach say it will encourage the development of alternative energy sources and energy conservation while maintaining the state's tax revenue (Christian Democrats, 1993). Although this approach looks appealing it has several problems. Firstly, the state cannot tolerate reduced revenues, so if an environmental problem is solved through tax switching, another source of taxation has to be found (Conservatives, 1995). Moving from one environmental problem to another increases the cost of the programme. Secondly, there is the problem of who should set up and monitor the taxes. If, for instance, a tax is placed on electricity consumption to alleviate the phase out of nuclear power many groups would protest. Energy intensive industry would cry foul play and that they were being forced into extinction, while the Conservatives would accuse the antinuclear parties (Centre and Communist) of trying to destroy Sweden's economy. However, these problems aside, tax switching could lead to significant CO_2 reductions.

Renewable Energies

Over the past 20 years the Swedish government has invested 2.1 billion SEK in developing biomass, 1.4 billion SEK in solar power and 976 million SEK in wind power (Holm, 1995). However, little research has been done to determine whether these investments have given value for money. Biomass and peat together account for 76 TWh of Sweden's overall energy mix which represents a considerable amount of energy, but it is unclear how much of this came about as a result of the cash injections. Some of the investments and the resulting CO_2 savings, particularly in the pulp and paper industry, would have taken place without government subsidies. The development of biomass fuelled district heating would appear to be the area that has gained most from the government money.

In 1991 the Government announced that (over a five year period) 625 million SEK (approximately 60 million pounds) would be allocated to build large scale biomass plants for electricity production. The money was administered by NUTEK, and three biomass plants were built, including one in the city of Växjö (discussed on pp178–179). Additionally, the Government set up the Biomass Commission to examine the future potential of biomass use in the domestic heating and electricity sectors. The Commission's findings showed that there was a real potential to increase biomass use from its current levels of approximately 70 TWh a year to up to 185–220 TWh if all sources (eg clear cut residues, sawdust from sawmills and energy forests) were included (Swedish State Studies, 1992).

The government subsidies for biomass and the CO_2 tax on fossil fuels have increased the popularity of biomass. The government believes that the CO_2 and other environmental taxes will encourage more and more utilities to convert their district heating plants from fossil fuels to biomass. Indeed, bioenergy increased from 43 TWh of the nation's energy mix in 1970 to 65 TWh by 1990 (NUTEK, 1991) and 78 TWh in 1994 (NUTEK, 1995).

By far the largest producer and user of biomass for energy purposes is the pulp industry. In 1994 this use represented 30 TWh (NUTEK, 1995). The use of biomass for district heating purposes has, however, increased dramatically over the past 15 years. From less than 1 TWh in 1980, wood fuels today account for 9.8 TWh of the total biomass used for heating purposes. In fact, between 1992 and 1994 the use of wood fuels in this sector almost doubled from 5.4 to 9.8 TWh due to favourable tax legislation (unlike fossil fuels, wood fuels are not subject to either CO_2 or SO_2 taxes) (NUTEK, 1995). Present trends indicate that there will be more conversions of oil to biomass-fuelled district heating boilers over the next few years, leading to reductions in CO_2 emissions.

Following the 1991 Energy Act, the use of wind power for electricity generation was actively encouraged by the government. In the period 1991–1996, the government allocated 250 million SEK to subsidize the capital costs of building windmills by 35 per cent, and as of 1st July 1994 electricity from windmills in Southern and Central Sweden received a special environmental bonus of 0.088 SEK per generated kWh (0.036 SEK per kWh in Northern Sweden) (NUTEK, 1994b). Owing to these large subsidies, wind power is becoming a popular energy alternative in Sweden, although it is highly unlikely that it will contribute significantly to the overall energy mix. In 1994 wind power accounted for only 0.072 TWh of Sweden's total electricity production of 138 TWh, and by the year 2005 wind power will contribute no more than 0.2 TWh out of predicted total of 152.2 TWh based on current investment and production levels (NUTEK, 1994b).

In short, the overall success of government policy on renewables has been mixed. More systematic support for the energy sources that have the greatest potential for commercial viability is necessary if the government wants to achieve CO_2 reductions in this way. At present, biomass is the most logical choice.

Energy Conservation

Swedes are experts in the area of energy conservation, with the highest building standards of all OECD nations. There are two main reasons for this. Firstly, the housing stock is very new, with the majority of homes being constructed post 1964 as a result of the government's 'one million' house programme. These homes were built to surpass Sweden's already strict building codes as the builders believed (rightly) that the government would make the codes even stricter as time went on (Schipper et al, 1985). Secondly, as a result of the amount of energy conservation legislation introduced in the 1970s, which enabled home owners to obtain free building inspections and subsidized retrofitting, most homes have very high thermal standards (Schipper et al, 1985; Vedung, 1982).

NUTEK's board on the Effective Use of Electricity is the main body looking at energy efficiency in Sweden. It has a budget of 400 million SEK over five years (NUTEK, 1994c). Its specific purpose is to determine the potential for reducing Sweden's electricity use by 10 per cent by the year 2000, but also more generally to identify areas where energy conservation may lead to reductions in CO_2 emissions. The board has looked at encouraging manufacturers to improve refrigerator technology, increasing the energy efficiency of windows and introducing DSM on a regional scale, among other measures (NUTEK, 1994c).

The Phase Out of Nuclear Power and the Energy Commission

In their 1994 election campaign the Social Democrats promised to phase out one nuclear reactor during their term in office as a way of winning votes from the antinuclear Centre and Green Parties. As a result, just as in 1991, the future of nuclear power is once again on the political agenda.

Upon taking up office, the Social Democrats assigned the cross-parliamentary Energy Commission, set up by the previous government, to look at a range of energy scenarios in the light of a potential phasing out of nuclear power, starting with one reactor before 1998. In December 1995 the findings of the Energy Commission were published (Swedish State Studies, 1995). They showed that it would be very difficult for Sweden to phase out nuclear power completely by the year 2010, but that it would be possible to phase out one nuclear reactor before the next general election (in 1998). The commission also felt that a phase out of one nuclear reactor should be encouraged as this would ease the transition into a new 'energy system' (Swedish State Studies, 1995). Following the publication of the Energy Commission's findings, the government has set up a political energy policy group to decide the future of Sweden's nuclear reactors. This group will not make a decision until early 1997, but already there are clear signals that if a nuclear phase out is begun in the short term, then nuclear generated electricity will have to be replaced by natural gas, leading to increased CO_2 emissions.

The uncertainties surrounding the future of nuclear power and the role of the market in Sweden have had important effects on the country's climate change strategy. If nuclear power is phased out CO_2 emissions will increase dramatically at home or in neighbouring countries (depending on whether Sweden imports electricity or natural gas and other fossil fuels to generate electricity domestically). Hence, if the government wants to phase out nuclear power it needs to address how it will deal with the increased CO_2 emissions. So far no policy indications have been forthcoming. The government made a pledge in 1980 to phase out nuclear power by 2010, but its CO_2 emission scenario for 2010 does not account for a nuclear phase out.

Based on the current political and economic situation, this author believes that the phase out of nuclear power will not occur. The main reason is that the country cannot afford it. As the government is in debt, movements in world financial markets have major effects on Swedish interest rates. If policy makers decide to phase out nuclear power, it is likely that world financial markets will react negatively. In the light of its high loan exposure Sweden cannot lose this relatively cheap energy source.

TRANSPORT ISSUES

Sweden is a sparsely populated country and most urban centres are far apart. As a result, Sweden's transport policy attempts to diminish its comparative disadvantage in this area. Sweden's road infrastructure is nowhere near full capacity. As a result of the emphasis on road transport, environmental issues have had little influence on transport policy. The car is seen as a 'holy cow', vital as a means of communication particularly in rural areas of Sweden.

The transport sector is the fastest growing source of CO_2 emissions in Sweden at the present time. Depending on which predictions one accepts, CO_2 emissions in the transport sector are set to increase by between 16 and 24 per cent by the year 2005 (Swedish State Studies, 1994). There are several reasons for this. Since 1950 the Swedes have increased their travel from an average of 10 km per day to 40 km, and it is thought that this will increase to 50 km by the year 2020 (Swedish State Studies, 1994). Since 1950 the number of vehicles and trucks on the road has grown exponentially. In 1950 there were 252,000 cars and 85,000 trucks, today there are 3.6 million cars and over 300,000 trucks (Larsson, 1995). Of the total freight in 1990 (455 million tonnes), 388 million tonnes was transported by road.

In the roads sector several major infrastructure projects have already been initiated. Examples are the planned Öresund bridge linking Malmö with Copenhagen and the so-called Dennis Package to build a ring road around Stockholm. Investments in the railways, the largest since 1860, will counteract some of the projected increase in road traffic. Currently several major modernization projects are underway, including major improvements to the main lines between Stockholm and Malmö and Stockholm and Gävle, which will cut travel times by several hours.

Finally, following the deregulation of air travel, domestic air traffic has increased dramatically. Linjeflyg was the sole domestic carrier (part of the Scandinavian Airline System network), but following deregulation in 1992 several other airlines have been set up. Of these, the largest is Transwede with domestic and international services. Deregulation has also led to the growth of Arlanda airport. As recently as the early 1980s there was only one terminal at Arlanda. Today there are four and the number of passengers has increased from 9.2 million in 1985 to 11.1 million in 1992. Deregulation may have led to cheaper airfares on competitive routes and better service, but the rise in air traffic has increased fossil fuel use, contributing to more CO_2 emissions in the transport sector. In summary, all types of traffic are set to increase. As shown in Table 10.3 (p168), the transport sector will contribute the most to the increased CO_2 budget.

The government has instigated several measures to address the increase in CO_2 emissions from the road traffic sector. The most significant of these has been successive tax rises on gasoline and diesel fuels. In 1990 CO_2 taxes added 10 per cent to the price of gasoline and diesel and in 1991 VAT increased the prices by an additional 25 per cent, resulting in some of the highest fuel costs in Europe. This reduced gasoline consumption temporarily. Between 1990 and 1991 gasoline use decreased by 8 per cent and although it increased in 1992 by 5 per cent, it decreased again by 2 per cent in 1993, but in 1994 it increased by 3 per cent (NUTEK, 1994b; 1995). The taxes on gasoline are seen by the government as an important strategy to combat CO_2 emissions. Reductions of 2.2 million tonnes of CO_2 based on 1990 levels are projected by the year 2000 (Swedish Ministry of the

Environment and Natural Resources, 1994a, 1994b).

Policy makers are also in the process of introducing more environmentally friendly local transport. The head of the Stockholm Road Administration, Hans Rode, has outlined several strategies:

■ a pilot scheme to introduce ethanol fuelled city buses;
■ the construction of a ring road around Stockholm (the Dennis Package) to reduce pollution levels in the city;
■ an 'integrated travel package' bringing together road, underground and local railways to reduce the environmental impacts of commuting. Stockholm already has a 300 SEK monthly pass which allows city wide travel on any form of public transport. This represents very good value for money.

The government has also initiated a series of research programmes on the effects of traffic on the environment. The project most directly relevant to the subject of this publication is the Traffic and Climate Committee who are looking at stabilizing CO_2 emissions in the sector by the year 2000 based on 1990 levels. The Committee has suggested a five pronged approach consisting of the following :

■ exemption from CO_2 taxes for biofuels used for transportation (eg ethanol);
■ reduced sales taxes on cars with low fuel consumption;
■ stricter controls on fuel use in different types of engines to ensure appropriate fuels are used, reducing pollution problems;
■ levying a uniform CO_2 tax on all vehicles;
■ increasing the competitiveness of public transport (Swedish State Studies, 1994).

More could be done, however, as there is considerable potential for energy conservation in the transport sector. Sweden's cars are on the whole larger, heavier and have more engine power than cars driven in the rest of Europe (Schipper and Meyers, 1992; Schipper et al, 1994). The two main domestic car manufacturers, SAAB and Volvo, which dominate the Swedish market, specialize in large, high performance vehicles and do not produce small energy efficient cars. In addition, company cars, which generally fall within this category, account for 10 per cent of the car market (Schipper et al, 1994). Little research has been done on reducing fuel consumption in cars. This is clearly seen in examining data from average fuel consumption for new cars in Sweden. In the period 1978–1987 new cars became more fuel efficient with a decrease in fuel use from 9.3 to 8.2 litres per 100 km. However, after 1987 this trend halted and reversed. In 1994, for example, the average fuel consumption for a new car was 8.3 litres per 100 km, 0.1 litres higher than the figure for 1987 (NUTEK, 1995).

The situation is no better in the truck sector. Per tonne km energy intensity in trucks has increased mainly due to the greater use of light trucks, which per km and weight transported are more energy intensive than larger ones. Additionally, there have been only small efficiency improvements in truck engines over the past 30 years (Schipper et al, 1994).

ACTIONS IMPLEMENTED AT THE LOCAL LEVEL

On the whole local government has been heavily involved in promoting an environmentally sustainable society. In Sweden environmental policy is driven by the Ministry of the Environment and Natural Resources. It develops environmental policies such as CO_2 taxes, on which local government can then develop their own policies.

Local governments obtain funds for environmental improvements in three ways. Firstly, they can take an environmental levy from municipal taxes. With this money the municipalities fund environmental advisors for the city to develop environmental campaigns, such as to encourage citizens to save energy. Secondly, most municipalities have their own energy company and some of the revenue from this can be set aside for energy efficiency schemes and investments in renewable energy sources. Finally, municipalities receive advice and at times finance directly from central government.

One area where local governments have been particularly active is in developing Local Agenda 21. In fact, implementation of Agenda 21 is seen as a major success for the former Conservative government (Bjerström, 1994). To see how the Local Agenda 21 initiatives worked in practice, environmental and energy policy issues were discussed with three municipalities: Eskilstuna, Uppsala and Växjö. An important aspect of the 'greening' of municipalities has been the increasing use of biomass in district heating plants. This has reduced fossil fuel consumption which has led to less CO_2 being emitted. All three of the municipalities investigated had some form of environmental and energy strategy in place. However, one must not forget that the standards for environment and energy are already among the highest in the world (Flam and Jamison, 1989). Three pane windows are the norm and 50 cm of insulation in houses is usual. This is augmented by economic mechanisms, such as the CO_2 and SO_2 taxes, and the energy planning advice made available by NUTEK.

As a result of these financial and regulatory mechanisms, all three municipalities visited have adopted strict energy conservation measures in their respective administrations (eg installing passive lighting systems and low energy light bulbs in municipal buildings) and have promoted domestic energy conservation. The municipalities visited have also developed extensive district heating networks fuelled, to some extent, by alternative energy sources.

Eskilstuna, a medium sized town with a population of 60,000, is located 140 km south west of Stockholm. The city was until recently the only town in Sweden which had joined the so-called Climate Alliance, calling for a CO_2 reduction of 50 per cent by the year 2010. In this sense the city was seen as pioneering in the field of climate policy. However, local officials readily admitted that they signed the agreement after it was known that the commitment could be met by installing a new wood chip boiler at the district heating plant. Investment in this type of boiler was commercially viable due to the environmental taxes levied on fossil fuel use (CO_2 and sulphur). The Agenda 21 officer in Eskilstuna did believe that more innovative environmental initiatives could be undertaken, such as bicycle paths and windmills, but these were only at the planning stage (Dahlin, 1995).

Uppsala, Sweden's largest university town with a population of 110,000, is located approximately 70 km north of Stockholm. As with other municipalities it has developed a series of environmental strategies but it has experienced difficulties. The district heating system is almost solely based on peat which is transported from the distant western province of Härjedalen.[4] Additionally, budgetary problems have led to a 25 per cent cut in the environmental budget.

Växjö, with a population of 70,067, lies about 450 km southwest of Stockholm. Of the municipalities reviewed for this study, Växjö has the most impressive environmental record. In 1980 the municipality was the first in Sweden to convert an oil fired boiler at its district heating plant to run on biomass. The conversion was so successful that the funding agency used the Växjö model for 25 other boilers throughout Sweden. The municipal energy company (VEAB) has since converted another boiler, enabling the municipality to meet 80 per cent of its heating needs from biomass. This will increase to nearly 100 per cent following a decision to build a new cogeneration plant at the district heating site. It is envisaged that the new facility will increase VEAB's electricity production from 17 to 38 per cent of Växjö's total consumption. This will reduce VEAB's oil requirement from 12 to only 2 tonnes a year, leaving the use of oil only for ignition and reserve purposes. In January 1995 the city became a member of the Climate Alliance, committing itself to a CO_2 reduction of 50 per cent by the year 2010.

CONCLUDING COMMENTS

Sweden has built up an international reputation for its progressive environmental policy (Flam and Jamison, 1989). However, in the area of climate change the signs are that the present government is less than enthusiastic about pushing domestic CO_2 abatement policies. As a result, the political feasibility of further actions beyond revenue generators such as the CO_2 tax is uncertain.

At the moment, climate change is not high on the political agenda. Although the government has raised industry's CO_2 tax somewhat, it had already cancelled NUTEK's very successful pilot programme of joint implementation, cut the Environment Ministry's annual budget by 18 per cent (more than for any other Ministry) (Olsson, 1994) and phased out the Swedish National Committee for Climate Change by the end of 1995 (Daleus, 1995). The government is not addressing the major source of CO_2 emissions

---

4 Although peat is seen as an alternative fuel, it is debatable whether or not it is renewable or if it adds any net CO_2 emissions to the atmosphere.

in Sweden, the transport sector. The phasing out of nuclear power, if it goes ahead, would indirectly increase CO_2 emissions because the country would become increasingly reliant on fossil fuels for its energy supply.

The Government's stance is by no means new. In 1990, when the current administration was last in power, it was criticized at the World Meteorological Organization Climate Conference for pushing for international policy agreements without formulating domestic policies (Klefbom, 1993). National CO_2 emissions have increased by 5 per cent since 1990, with current indications showing that Sweden will not be able to stabilize CO_2 emissions by the year 2000.

There are several reasons for this domestic inaction on climate change. Firstly, there is the 'feel good' factor among Swedish politicians. Sweden reduced its CO_2 emissions by 40 per cent between 1970 and 1990 and it is one of the few nations to implement a CO_2 tax. Hence, policy makers today believe that further cuts in the country's CO_2 emissions are unwarranted compared to the need in other countries. In a recent interview, the Environment Minister, Anna Lindh, stated that Sweden is not a main player in climate change because it began to reduce CO_2 emissions much earlier than anyone else (Eneberg, 1995). Secondly, in contrast to environmental problems such as acid rain, Sweden is unlikely to be adversely affected by global warming. In fact, it may benefit with warmer winters and a longer growing season. Thirdly, the per capita costs of stabilizing or reducing CO_2 emissions will be incredibly high for Sweden. In a recent nine country comparison, the so called MARKAL model suggested that to stabilize CO_2 emissions by the year 2020 Sweden would have the highest marginal costs except for Norway (Hill and Kamp, 1994). Policy makers, aware of Sweden's budget problems, do not want to inflict this on the economy.

Also, Sweden has a large forest sink, with half the total CO_2 emissions being absorbed by the nation's forest. Hence, some industrialists and policy makers feel that no dramatic CO_2 abatement measures are needed (Lodin, 1995) due to this temporary CO_2 absorption cushion.

To reduce CO_2 emissions, cars need to be more fuel efficient and people must be encouraged to drive less. This, however, will not happen as the transport sector is seen as an untouchable necessity and an inviolable right to citizens in a country with such a low population density. The nuclear power debate further complicates projections for CO_2 production levels. If it is to be phased out politicians cannot afford to be concurrently campaigning for a stabilization or even a reduction of CO_2 emissions, since doing so could easily result in their being singled out for criticism as proponents of environmental protection no matter what the detrimental effects might be upon economic prosperity or stability. In short, this could mean political suicide.

It is clear, therefore, that current measures will not be able to sufficiently curb the growth of CO_2 emissions in Sweden in the short term. Until it addresses the uncertainty posed by the potential phase out of nuclear power, Sweden needs to implement other CO_2 abatement measures if it wishes to meet its climate convention commitments. As it is difficult to reduce CO_2 emissions domestically without problems for the economy, it would make sense for the country to become active in joint implementation and then instigate a tradeable emissions permit system in the longer term. However, the current government is moving the country away from being at the forefront of joint implementation programmes and research to being a minor actor.

REFERENCES

Agricultural Ministry (1993) *Jordbruket och Skogsbruket som Resurs i Klimatarbetet*, Agricultural Ministry, Stockholm

Bjerström, E (1994) 'Borgerlig oenighet försämrade miljön', *Dagens Nyheter*, 25th July 1994, pA7

Christian Democrats (1993) *Skatteväxling för miljöns skull*, Christian Democrats, Report No 13, Stockholm

Conservatives (1995) *Motion till Riksdagen 1994/95 m509*, The Conservative Party, Stockholm

Dahlin, L E (1995) Agenda 21 Officer for Eskilstuna municipality, personal communication, April

Daleus, L (1995) *Riksdagen Protocoll*. 1994/1995:76, 21, State Documents, Stockholm

Eneberg, K A (1995) 'Jag har gett mitt bidrag', *Dagens Nyheter*, 30th March, pA10

Eriksson, L (1991) 'Biobränslet vinnaren', *Nyteknik*, No 4

Flam, H and Jamison, A (1989) 'The Swedish confrontation over nuclear energy: a case of timid antinuclear opposition', In Swedish Collegium of Advanced Study (ed) *SCASS Study of Antinuclear Movements and the State*, Swedish Collegium of Advanced Study, Uppsala

Hill, D and Kamp, T (1994) 'Boundaries of future carbon dioxide emission reduction in nine industrial countries', *Change 21*, August

Holm, M (1995) 'För tidigt att fatta nytt energibeslut', Miljöaktuellt, 1st March, p4

Jordbruksutskottet (1987–1988) *Miljöpolitiken-prop. 1987/1988:85*, Swedish State Publications, Stockholm

Kågeson, P (1990a) 'Uppgörelse på väg om energin', *Dagens Nyheter*, 8th June, pA4

Kågeson, P (1990b) 'Lås er inte vid 2010!', *Dagens Nyheter*, 11th October, pA4

Kempton, W (1991) 'Lay perceptions on global climate change', *Global Environmental Change* Vol 1, no 3, pp183–208

Klefbom, E (1993) *Hejdlös Forskning: Växthuseffektens och Ozonförstöringens Politiska Sprängkraft*, Unpublished BA Thesis, Department of Government, Uppsala University, Uppsala

Larsson, A (1995) Letter from A Larsson at the Swedish Central Bureau of Statistics, May, Stockholm

Lodin, S O (1995) 'En miljöfarlig miljöskatt', *Dagens Nyheter*, 6th February, pA4

Löfstedt, R E (1991) 'Climate change perceptions and energy use decisions in northern Sweden', *Global Environmental Change* Vol 1, no 4, pp321–324

Löfstedt, R E (1992) 'Lay perspectives of global climate change in Sweden', *Energy and Environment* Vol 3, no 2, pp161–175

Löfstedt, R E (1993a) *Dilemma of Swedish Energy Policy*, Avebury, Aldershot

Löfstedt, R E (1993b) 'Risk communication in the Swedish energy sector', *Energy Policy* Vol 21, no 7, pp768–772

Löfstedt, R E (1993c) 'Lay perspectives concerning global climate change in Vienna, Austria', *Energy and Environment* Vol 4, no 2, pp140–154

NUTEK (1991) *Energiläget 1991*, NUTEK, Stockholm

NUTEK (1994a) *Energiläget 1994*, NUTEK, Stockholm

NUTEK (1994b) *Klimatrapporten*, NUTEK, Stockholm

NUTEK (1994c) *Programmet för Effektivare Energianvändning*, NUTEK, Stockholm

NUTEK (1995) *Energiläget 1995*, NUTEK, Stockholm

Olsson, H (1994) 'Hård besparing på miljön', *Dagens Nyheter*, 17th November, pA11

Schipper, L, Johnsson, F, Howarth R, Andersson, B and Price, L (1994) *Energianvändningen i Sverige: ett Internationellt Perspektiv*, NUTEK, Stockholm

Schipper, L and Meyers, S (1992) *Energy Efficiency and Human Activity: Past Trends and Future Prospects*, Cambridge University Press, Cambridge

Schipper, L, Meyers, S and Kelly, H (1985) *Coming in from the Cold – Energy Wise Housing in Sweden*, Seven Locks Press, Washington DC

SNCC and Swedish EPA (1995) *Analys av Frågor i Anslutning till Klimatkonventionen*, Swedish Environmental Protection Agency, Stockholm

Swedish Bureau of Statistics (1994) *Utsläpp till luft i Sverige av Koldioxid 1993*, Swedish Bureau of Statistics, Stockholm

Swedish Bureau of Statistics (1995) *Utsläpp till luft i Sverige av Koldioxid 1994*, Swedish Bureau of Statistics, Stockholm

Swedish Energy Administration (1989) *Ett Miljöanpassat Energisystem*, Liber Förlag, Stockholm

Swedish EPA (1995) *Sverige mot Minskad Klimatpåverkan*, Swedish Environmental Protection Agency, Stockholm

Swedish Heating Association (1994) *Statistik 1993*, Swedish Heating Association, Stockholm

Swedish Ministry of the Environment and Natural Resources (1994a) *Sweden's National Report: Under the United Nations Framework Convention on Climate Change*, Swedish Ministry of the Environment and Natural Resources, Stockholm

Swedish Ministry of the Environment and Natural Resources (1994b) *Så Fungerar Miljöskatter!*, Fritzes, Stockholm

Swedish Parliament (1987–1988) *Regeringens Proposition: 85 Miljöpolitiken inför 90 talet*, State Documents, Stockholm

Swedish Parliament (1989–1990) *Regerings Proposition: 111 om koldioxidskatten*, State Documents, Stockholm

Swedish Parliament (1990–1991) *Riksdagens Proposition: 88 om Energipolitiken*, State Documents, Stockholm

Swedish Parliament (1992/93) *Regeringens Proposition 179: Atgärder mot klimatpåverkan m.m*, State Documents, Stockholm.

Swedish State Studies (1992) *Biobränslen för Framtiden: Slutbetänkande av Biobränslekommissionen*, Allmäna Förlaget, Stockholm

Swedish State Studies (1994) *Trafiken och Koldioxiden: Principer för att Minska Trafikens Koldioxidutsläpp*, Fritzes, Stockholm

Swedish State Studies (1995) *Omställning av Energisystemet: Slutbetänkande av Energikommissionen*, Fritzes, Stockholm

Vedung, E (1982) *Energipolitiska Utvärderingar 1973–81*, Liber Förlag, Stockholm

WMO (1986) *Report of the International Conference on the Assessment of the Role of Carbon Dioxide and of Other Greenhouse Gases in Climate Variations and Associated Impacts*, Villach, Austria, 9th to 15th October 1985, WMO No 661

World Commission on Environment and Development (1987) *Our Common Future*, Oxford University Press, Oxford

Chapter 11 | COMPARATIVE ANALYSIS AND CONCLUSIONS

Ute Collier and Ragnar E Löfstedt

INTRODUCTION

The analysis of the case studies reveals, as was to be expected, a host of different attitudes and developments with regards to the climate change issue. Yet, one common trend found in five out of the six countries is that emission reductions are progressing at a slower pace than expected or that emissions are even projected to increase. This indicates that current policies are inadequate and puts into question the realization of most emission targets. Problems with reaching emission targets are to some extent, to be expected in those countries where environmental policy generally does not have a high priority (such as France, Italy or Spain). In fact, Spain has always projected emission increases, arguing, along with other cohesion countries, that these were inevitable as its economy needed to catch up with the rest of the EU. However, even proactive countries such as Germany and Sweden are encountering problems. In Sweden stabilization looks unlikely, with the government fairly unconcerned about the situation, while in Germany emission reductions are likely to slow as soon as the new *Länder* emerge from the post unification economic slump. The proactive countries have been quick to set emission reduction targets (in the case of Germany a very ambitious one) but their implementation is not proving easy. The aim of the following sections is to comparatively evaluate the main constraints to realizing emission reductions in the case study countries.

COMPARING THE NATIONAL PROGRAMMES

While the introduction has already raised some doubts about the effectiveness of individual countries' responses to the climate change issue, it has to be recognized that all have become engaged in drawing up climate change policies. In some countries (France, Italy, Spain and the UK) this process was set in motion after the signature of the FCCC, while Germany and Sweden took unilateral action before this. However, emission trends are not necessarily a very good guide to how active and committed countries are. The UK's climate change strategy, for example, is essentially defunct and its government is relying almost entirely on beneficial incidental effects of electricity sector privatization. While incidental benefits are in general to be welcomed, the problem in this case is that they are likely to be short lived.

When comparing the six countries it is difficult to make any kind of classification with regard to the notion of policy 'leaders' and policy 'laggards', independent of the actual emission trends. Sweden shows some commitment to action, having drawn up a comprehensive strategy and having made some progress towards implementing this (eg through the application of carbon taxes). Germany likes to project itself as a leader country (eg at the Berlin Conference of the Parties) but has so far failed to implement the majority of the proposals made in its own climate strategy. In particular, over the past year climate change has descended towards the bottom of the political agenda. Meanwhile, none of the other four countries examined herein distinguishes itself in terms of either comprehensive programmes or effective measures, relying mainly on the incidental benefits of measures implemented for other reasons.

A number of general observations can be made as regards policy developments. Capacity building, ie the building up of knowledge, as well as of new institutions to deal with climate change, has been an important aspect of the initial response to the issue. This was less important in Sweden, which already had elaborated environmentally focused energy programmes, to which their climate strategies became tightly linked. In Germany, the Enquete Commission, which produced a number of authoritative studies, became an influential actor in the policy process and policy coordination between the different ministries became increasingly important. Meanwhile, in France an Interministerial Committee was formed to deal with the issue, also aiming at policy coordination. In Italy, Spain and the UK, elaboration of the climate strategy was the responsibility of the respective environment ministries, with more or less strong consultation with other ministries. However, as later sections show, the integration of climate change concerns into other policies still remains minimal in most cases.

As far as the content of the actual climate change strategies is concerned, while there are some variations between the six countries, some common characteristics can be observed. In terms of the policy measures listed in the programmes, a common pattern of regulations and financial incentives, especially for energy efficiency and renewable energy, emerges. Not all the measures in the national programmes have been implemented as planned.

Table 11.1 shows a summary of the main measures currently being implemented, also indicating likely emission developments until 2000 (according to government projections). Furthermore, some comments on the country specific situation are included in the table.

Table 11.1. *Main Features of National Climate Change Programmes*

Country (projections for 2000)	Main measures currently being implemented	Comments
Germany (–10%)	Voluntary agreements with industry, efficiency improvements in new *Länder*, subsidies for renewables, stricter building standards, local authority and energy company actions	Main reliance on side effects from unification, continuation of existing measures
UK (reduction of up to 7%)	Energy efficiency programmes (EST), VAT on fuel, surcharge for renewables (NFFO), voluntary energy efficiency commitments in public and private sector	Main reliance on 'dash for gas', energy efficiency underfunded, falling energy prices compensating for VAT
Italy (+3%)	Energy efficiency subsidies (law n. 10/91), grants for CHP, some subsidies for renewables	Continuation of measures of 1988 energy plan, implementation problems due to lack of finance
France (+13%)	Subsidies for renewable energies, reforestation, public transport subsidies, some incentives for energy efficiency	Reliance on past achievements, limited scope for emission reductions in electricity generation
Spain (+15%)	Support for CHP, small energy efficiency programme, subsidies for renewables (plus use of EU structural funds)	Continuation of measures under 1991 national energy plan, cut back for energy efficiency measures
Sweden (increase between 6 and 15%)	CO_2 tax, energy efficiency programmes, subsidies for biomass plants	Dilemma with nuclear phase out, large previous reductions

As the third column of Table 11.1 shows, many of these measures are simply a continuation of already existing policies, generally conceived to reduce oil dependence and to improve energy efficiency, but now dressed up as climate change instruments. However, climate change certainly has given a new incentive to energy efficiency programmes which, because of the low concern about energy prices and security of supply, had decreased in significance. Some new measures have been employed, although none of these are particularly radical. Furthermore, there are clear tensions with other policy objectives, such as the pursuit of low energy prices for industrial competitiveness.

The case studies also reveal that the application of fiscal measures to reduce emissions is not widespread and is fraught with difficulties, despite a considerable interest in the potential benefits of using economic instruments. In Sweden and the UK, fiscal measures have been weakened. In the case of Sweden, this has been due to industrial opposition, in the case of the UK mainly because the opposition parties forced the issue off the agenda. The UK tax (VAT on domestic fuel) was conceived on economic grounds but subsequently dressed up as an environmental measure. The provision of compensatory measures, which could have been expected to reduce economic constraints, has not improved political feasibility. Elsewhere, energy prices show a downward trend both because of low world market prices and regulatory decisions. Only in Italy have energy prices continued relatively unchallenged at a high level as a result of government taxation, based on economic grounds. This has had a beneficial impact on energy efficiency, for example in the car fleet. However, as income levels have grown substantially in recent years in Italy, energy bills have become a smaller component of household expenditures so incentives for energy efficiency have decreased.

In the cases of Germany and the UK, voluntary agreements are supposed to play an important role in achieving emission reductions. However, in neither country are these agreements given a legal standing. It is thus not clear how effective such agreements will be without the possibility of legal enforcement. In general, the studies show that there has been much talk about voluntary agreements (also in France and Spain) but, in reality, little progress has been made.

The programmes are particularly weak as far as measures for the transport sector are concerned. The UK programme, for example, proposes a campaign to change driving behaviour as a way of influencing fuel efficiencies of vehicles (reducing CO_2 emissions by encouraging lower speeds and less aggressive acceleration, which is unlikely to be very effective), while the Italian and Spanish programmes ignore transport altogether. Hence, the problems encountered in the meeting of emission targets are partly a result of the inadequacy of the climate change strategies themselves and partly due to a failure to implement the proposed policy measures. There are a number of explanatory factors for these problems, which are dealt with in the following sections.

THE INFLUENCE OF EMISSION CHARACTERISTICS

There is no single factor (even for individual countries) which can explain the difficulties in meeting emission targets. As already mentioned, the predominant fuel used, especially in the electricity sector, can be expected to play a role. Those countries which currently use a high level of coal have available, in principle, relatively 'easy' opportunities for emission reductions because of the possibility of fuel switching (from coal to natural gas, nuclear or renewables). In particular the switch from coal to natural gas in electricity generation is currently also economically attractive, as the examples of Germany (especially in the new *Länder*), Italy and the UK show. France, where over 90 per cent of electricity is generated from non-fossil sources, in particular nuclear, has almost no scope for fuel switching.

The situation is similar in Sweden, where nuclear and hydropower dominate. As in France, CO_2 emissions in Sweden saw a substantial reduction prior to 1990, a fact which these two countries argue should be taken into account when assessing their current policies. In Sweden, the situation is further complicated by the fact that a phase out of nuclear power was decided in a public referendum. As a direct consequence of this, the role of fossil fuels in electricity generation is likely to increase rather than decrease. Additionally, Sweden has in the past pursued an aggressive energy efficiency policy, so there is a limited potential for emission reductions in this area because of absolute technical constraints.

Meanwhile, in Germany and Spain fuel switching is in reality quite difficult because of the availability of indigenous coal reserves. Reducing coal use in these countries is both an economic (balance of payments) and social (employment in coal mining) cost and thus has political implications. In principle this also applies to the UK, yet here substantial emission reductions are resulting from a major fuel switch from coal to natural gas. This is occurring not as a result of specific climate change policies but as a byproduct of privatization and liberalization in the electricity sector. While there has been political opposition to the destruction of the coal industry, this has been ineffective. This shows clearly that governments can overcome such constraints if they have enough political determination and, as in this case, doggedly follow their ideological aims in the face of all opposition. The justification for policies employed in these cases is economic rather than environmental and hence likely to be more acceptable to those on whom the present government can expect to rely for continued political support.

In general, countries which have little scope for reductions from the electricity sector have to rely on emission reductions from other sectors, in particular transport, industry and the domestic and/or commercial sectors. As the case studies show, the transport sector is absolutely crucial, as it is here where the fastest growth in emissions is occurring. At the same time, it is a sector in which emission reductions are particularly difficult to achieve. While technical solutions (such as more efficient cars and lorries) can play a role, a substantial shift from individual to public transport is fundamental to reversing emission trends. Such a move will incur substantial investment costs and may need changes in people's mobility requirements. Both economic and political constraints are thus important in this sector.

While the above factors can play a role, the analysis of differences related to the shares of different fuels in energy consumption cannot provide a sufficient explanation for the differences between the case study countries, or for the more general problems associated with achieving emission reductions. The following section identifies a number of other important constraints.

POLITICAL CONSTRAINTS

The previous section has already hinted at a number of constraints which are impeding the achievement of emission targets, apart from the fuel share related factors. These can generally be described as political constraints, as opposed to technical and economic constraints, although these three types of constraint often interact. Some are clearly particular to individual countries. One example here would be the Italian case of failing to implement a number of important laws (laws 9 and 10 on energy efficiency and renewable financing, as well as the law requiring the drawing up of urban energy plans). This is indicative of a general implementation problem, political instability and severe budget constraints. Another specifically Italian problem is the scale of diffusion of responsibility for transport planning and policy between no less than 21 public authorities, which creates some confusion. However, cooperation problems between different levels of government are a more generic constraint.

Another very country specific problem occurred in Germany, where both political and public attention was distracted from the climate change issue as a result of unification. This attention resulted from the high economic costs associated with unification, but also the scale of other environmental problems to be dealt with in the former GDR. Furthermore, the speed of many decisions taken (such as electricity sector restructuring) has not always allowed consideration of all the important factors.

While some problems have thus been unique to a particular state, many of the constraints identified, while not necessarily being universally applicable, tend to recur in a number of cases. Among these are the following:

- Overriding priority allotted to economic growth and industrial competitiveness, linked with short term thinking.
- The lack of a consistent integration between environmental and other policies.
- Procedural constraints.
- Lack of cooperation between central and regional or local government.
- Institutional and regulatory frameworks for the energy sector.
- Low energy prices.
- The influence of the industrial lobby (especially with respect to tax proposals).
- Uncertainties about the role of nuclear power.
- Rapid growth in transport demand, eg due to trade liberalization and deregulation in the transport sector.
- Inherent attractiveness of private over public transport.
- Lack of public and/or pressure group interest.

Some of these constraints are interlinked. The following sections discuss these in turn.

Economic Priorities and the Lack of Policy Integration

The governments in the case study countries all claim to have recognized the importance of the climate change issue. Yet it is important to recognize that climate change concerns, together with other environmental issues, are almost always subordinate to overall economic policy, especially in relation to objectives such as economic growth and industrial competitiveness. These two areas are overriding policy priorities, whether in environmental 'leader' countries such as Sweden or 'laggards' such as Spain.

One problem within this context is certainly linked to the scientific uncertainties which continue to surround the climate change issue, as well as the long term nature of any likely effects. Climate change is thus often pushed aside by short term, seemingly more urgent priorities. However, as a number of economic models and analyses have shown, many measures which reduce greenhouse gas emissions can also make economic sense, and should thus be attractive in their own right. The case study analysis has confirmed this by revealing that emission reductions are occurring in a number of areas on account of decisions made for primarily economic reasons. However, there are many other examples in which measures to reduce emissions would be preferable in economic terms (such as those promoting energy efficiency) and yet are not being pursued. At least in Germany and Sweden there is discussion as to whether efforts to reduce CO_2 emissions could provide opportunities for technological innovation and changes in outdated industrial structures. As yet, the existence of such opportunities is not universally accepted and governments are generally not inclined to take what they see as risks.

One of the main obstacles to effective policy formulation and execution is that the economic benefits often occur only in the long term, while many governmental (and industrial) decisions are made with a short term horizon. Another problem is that while there may be aggregate economic benefits, particular sectors and individuals may lose out, resulting in considerable opposition to some measures. Furthermore, much decision making relevant to climate change occurs at the sectoral level (especially in the areas of energy and transport policies) where policy tensions can occur between different priorities.

The need to integrate environmental concerns in general, and climate change concerns more specifically, into areas such as energy and transport policy has received increased recognition both at the EU level and within the Member States in recent years. This line of thought follows the generally accepted principles that define the concept of sustainable development, which has been much discussed as a way forward in environmental philosophy and practice, particularly during and since the Rio summit. However, real policy integration has made limited progress to date, as recognized by the European Commission's Review of the Fifth Environmental Action Programme. Generally, environmental considerations continue to be addressed through marginal adjustments.

Procedural Problems

Part of the problem is procedural, with a lack of communication between different ministries. In Spain the elaboration of the climate change programme was the first time environment, energy and transport ministries had consulted each other. In Italy the transport ministry continues to make decisions without taking climate change concerns into account. As far as energy policy in Italy is concerned, the situation has improved somewhat since the 1987 referendum and strong local opposition to power plants has forced the integration of certain environmental concerns into energy policy, with some benefits in terms of greenhouse gas emissions. The main problem in Italy as far as energy policy is concerned is not integration as such, but rather implementation of the measures.

In France, while an interministerial committee was set up to elaborate the climate change strategy, normally energy and transport ministries show little concern for environmental issues. In Germany, there is much environmental rhetoric in policy statements from these ministries but other priorities continue to dominate. In the UK, each ministry has one official responsible for environmental matters but there is little evidence of real integration. One problem is that energy policy is the responsibility of either economic or industry ministries who traditionally have other priorities and are also heavy influenced by the industrial lobby.

In the UK the move of the EEO into the environment ministry (rather than the Department of Trade and Industry) has provided some integration but possibly simultaneously reduced the importance attached to energy efficiency due to the weak standing of the Department of the Environment. The low standing of environment ministries in the ministerial hierarchy is a problem that has also been found in a number of other studies on the integration of environmental concerns into other policies. Sweden had probably proceeded furthest along the integration path, at least as far as energy and environment policies are concerned, through the creation of a ministry that deals with both issues. However, the ministries were subsequently split again. There is nevertheless much policy coordination, in particular in relation to dealing with the nuclear dilemma.

In general, combined ministries are not necessarily a step forward. With a separate environment ministry, a voice for the environment is ensured, while combined ministries often subordinate environmental issues to other ministerial responsibilities. This is seen in countries like Spain and the UK, where environmental issues are part of much larger ministries dealing with other tasks such as public works or housing. In all countries, integration is most obviously lacking in the transport area, with hardly any attention being paid to climate change concerns in transport policy. A way forward might be the use of strategic environmental assessment procedures, but these have received little attention in the Member States examined.

Other Procedural Constraints and Lack of Cooperation

Procedural constraints have also been encountered in the case of specific policy instruments, especially voluntary agreements. In the UK, the lack of experience with a more consensual policy style has hampered progress. Rather than negotiating with industry, the government has drawn up a programme (the Making a Corporate Commitment Campaign) with some general guidelines, but has left the rest to industry, including the setting of targets. The agenda appears to be firmly set by industry, rather than by government policy priorities. A lack of experience with institutional cooperation between government and industry is also observed in Spain.

A further problem encountered is the lack of cooperation and coordination between different levels of government. As has been shown in the case studies, there is, in principle, much scope for action at the local (and in some countries regional) level and a number of local authorities have taken various initiatives. However, central governments have paid little attention to such initiatives when drawing up national climate change programmes. Meanwhile, local initiatives are hampered by unfavourable policy contexts and budget cuts. There is a need for better cooperation to avoid the duplication of efforts and to pool resources, as well as coordination to create a better awareness of the role of local and regional authorities.

Energy Sector Frameworks

The institutional and regulatory frameworks of the energy sector are clearly very important for emissions, but decision makers have paid little or no attention to this fact when drawing up climate change strategies. Subsequently, specific features of these frameworks have presented major constraints to the implementation of certain measures. On the institutional side, structure and ownership have some major implications. Public monopoly companies make very different choices when compared with liberalized, private companies. Nationalized monopoly companies can generally be characterized by their supply side mentality and scant attention to DSM and renewable energies, as the examples of EdF in France and ENEL in Italy demonstrate.

Energy sector liberalization offers some promise with regard to changing this situation, especially as far as the generating side is concerned. Large scale coal and nuclear plants are not very attractive for private sector investment and preference is given to gas fired plants and cogeneration, both of which are also beneficial in emission terms. However, as the example of the UK shows, appropriate regulatory frameworks need to be created to reap the benefits of a more open energy market. Important options such as DSM and district heating will not benefit from market liberalization per se and a new set of constraints can easily result. At the same time, public companies can, under certain circumstances, provide an energy system centred upon these options, as in the case of the *Stadtwerke* in Germany.

Meanwhile, certain features of current regulatory systems for energy companies (public and private) have been found to have negative, if unintended, effects. An obvious example is price regulation which can undermine other policy measures, especially in the field of energy efficiency. Furthermore, the link between profits and sales found in most countries acts as a disincentive to investments, both with private companies and where public energy companies are forced to operate in a commercial climate. Examples from the US regulatory system within the framework of Integrated Resource Planning can provide some ideas of how to address these problems. However, such regulatory changes have to date not been seriously considered by any of the Member States. The proposed EU directive on rational resource planning could be an important catalyst in improving regulatory systems.

Energy Prices, Taxes and the Industrial Lobby

Another important constraint is found in low energy prices, which occur as a result both of developments on the international energy markets (especially fossil fuels) and of regulatory intervention. It is well documented that low energy prices provide a disincentive to energy efficiency. Furthermore, they make renewables less attractive economically. Within this context the issue of market failure and external costs is important, as governments have the opportunity to intervene (especially through energy taxes) to redress the balance. Out of the case study countries, Sweden has applied a specific carbon tax, while the UK has introduced a 'pseudo' tax (ie by imposing VAT on domestic fuels). Germany has not moved on taxes despite a long and protracted discussion, while France and Spain oppose such a measure outright. Italy has high energy taxes in any case. Undoubtedly, the main constraint in this area has been the vehement opposition by the industrial lobby, as well as to some extent by the consumer lobby.

Carbon and energy taxes do not necessarily have negative economic and social effects, yet the discussion has rarely been conducted on a rational level. The strength of the industrial lobby is based on its ability to argue on the basis of economic growth and industrial competitiveness, which relates back to the points made on page 190. Sweden has dealt with industrial opposition by applying carbon taxes mainly to domestic consumers, which obviously reduces the potential effectiveness of this measure.

Nuclear Power

Nuclear power and uncertainties about its role have influenced policy developments in some countries (although in different ways), notably in Sweden and Germany, as well as to some extent in Italy and Spain. In Sweden, the decision to phase out nuclear power partly jeopardizes achievement of the emission stabilization target, as at least in the short term the easiest way to replace nuclear plants would be through fossil fuel plants. The situation is exacerbated by the fact that Sweden has already run an aggressive energy efficiency policy, so that cost effective emission reductions are not easy to come by. At present, the speed and scale of the phase out is rather uncertain, resulting also in uncertainties regarding the climate change programme.

In Italy and Spain there have also been referenda on nuclear power and, as a result, an expanding use of fossil-fuels. However, there has been little discussion of the nuclear issue within the framework of climate change policy in these countries. Meanwhile, in Germany disagreements about nuclear power between different actors have resulted in the collapse of the energy consensus talks which is one reason for the relative inaction in the energy policy area. The nuclear issue also demonstrates the importance of not losing sight of other environmental objectives and it is debatable whether or not nuclear power should have a role in a sustainable energy policy.

Transport Issues

The transport sector emerged as a difficult problem area in all countries. This has been the sector in which emissions have grown the fastest in recent years. Trade liberalization and deregulation in certain areas (eg air traffic deregulation in Sweden and bus deregulation in the UK) have been contributing factors. The main problem is the dominance of road based transport, especially private cars and lorries, although air traffic is also an increasingly important emission source. While there are possibilities for modal shifts, these can be costly, although high costs are not considered an obstacle to many road infrastructure projects. A more fundamental constraint is inherent attractiveness of the private car over public transport, as it offers much more flexible personal mobility and freedom. As far as freight transport is concerned, flexibility is also an important argument in favour of using lorries instead of trains.

Some of this inherent attractiveness can be countered through innovative approaches to land use planning to reduce the need to travel. Also, other environmental problems, such as increasing congestion and air pollution, have resulted in growing opposition to the expansion of private transport in some countries, especially Germany and the UK. These issues are less important in less densely populated countries such as France and Sweden, or those with currently a low level of car ownership, such as Spain. In Italy some beneficial measures are being taken as a result of severe air pollution problems in a number of cities.

Overall, as already mentioned, there is lack of integration of climate change objectives into transport policy, as exemplified by the overriding priority allocated to investments in road infrastructure almost everywhere. Furthermore, there is a strong car lobby in most countries. In France, Germany, Italy and Sweden the car industry is an important economic sector and there is a great political reluctance to tackle car dependency.

Public Awareness

Public awareness and interest did not emerge as a strong constraint in the analysis, but nevertheless play a role. Public awareness of climate change is probably greatest in Sweden, where environmental issues remain fairly high on the public agenda. In Germany, awareness is relatively high but interest in this issue has decreased on a par with the increase in the economic problems associated with unification. At the same time, out of the less environmentally proactive countries the Spanish public has shown some concern because of the potentially negative effects of climate changes, especially as a result of the drought conditions experienced in recent years. However, this heightened concern has not been accompanied by government action. In France and Italy growing environmental awareness in recent years has mainly focused on other, especially local, issues. Neither NGOs nor the governments have done much to promote awareness of the issue. Meanwhile in the UK there appears to be much public confusion about the issue, despite government and pressure group campaigns.

OPPORTUNITIES AT THE LOCAL LEVEL

The local government level is clearly important for transport and planning policies, but also for the implementation of other policy measures crucial to the reduction of climate change emissions. As the case studies have shown, local climate change policies are in existence in a number of EU Member States, although they vary in scope and in terms of the constraints encountered, both between countries and within countries. A number of explanatory factors for these variations can be identified. Firstly, the tradition for local involvement in environmental management is important. Here, Germany and Sweden, with generally a higher importance attached to environmental issues, are clearly ahead. In both countries local authority environmental departments have been in existence for one or two decades, while in Italy and Spain they are still relatively rare. In the UK most authorities now have an environmental policy officer but in many cases they lack a core staff. As at the national level, capacity building is important.

Secondly, the actual scope for action in terms of local competences and reduction potentials are significant factors. The municipal ownership of energy and transport companies allows local authorities to exert considerable influence on emission developments within these two sectors. However, as the cases of Germany and Sweden show, such competences do not automatically ensure efficient reduction strategies even if targets have been set. The specific problems differ between the two countries. In Sweden, the problem is primarily one of a lack of potential. Much has been done already in terms of energy efficiency, standards are high and public transport systems efficient. There is still potential as far as renewable energies (especially biomass) are concerned and progress is being made in some cases. However, these are often the more costly type of measures.

In Germany financial considerations are becoming increasingly important as local authorities are facing budget constraints. Energy companies are important in that their profits are a significant component of council budgets yet, under the current regulatory system of price control, companies are in effect undermining their profits when they promote energy efficiency too vigorously. Local councils themselves can ill afford to provide subsidies for promoting energy efficiency measures or renewables. At the same time low energy prices, further exacerbated by the abolition of the coal levy, provide little incentive for investment by energy consumers in these options. Furthermore, providing efficient and affordable public transport systems can be a drain on resources.

Where there is no municipal ownership of energy companies, as in the UK, local authorities can still be influential in that they are important energy consumers themselves, for example in terms of streetlighting and public buildings. Here emissions can be reduced through investments in efficient technology, insulation and various energy management activities, which generally also allow major cost savings. However, such activities are often neglected.

Of the case study countries, it is only in the UK that there is no public ownership of local transport companies. However, even there local authorities have a pivotal role in managing both transport demand and supply. Generally, local authorities in the EU have some responsibilities for road and cycleway provision. Traffic management is also important, as improved vehicle performance in less congested conditions means fewer CO_2 emissions. Additionally, measures such as the introduction of bus lanes and the reduction of parking can encourage intermodal shifts, as they change the relative attractiveness of travel by different modes.

The examples discussed in the case studies effectively present the 'best practice' cases in each country. In some countries, such as Italy and Spain, action is very much centred on a very small number of proactive municipalities. But even in environmental 'leader' countries, such as Germany, not all local authorities are involved and only a relatively small number stand out as very active. The level of local involvement is often very dependent on specific personalities, either at the political level, in the local administration or in the local energy company.

One clear message from this comparative study is that local authorities, with the best will and reduction potential, cannot design and implement effective climate protection policies in the absence of a supportive national, as well as EU, framework. This concerns, in particular, issues such as adequate financing of the local authorities and energy prices. Budget constraints appear to be a major obstacle everywhere. Some measures, such as energy management in local authority properties, can actually generate financial resources which can be used elsewhere. However, others, such as subsidies for energy efficiency, renewables or public transport, can require substantial financial resources, which may simply not be available. As far as energy efficiency is concerned, Third Party Financing models can provide a way forward. Furthermore, where there are municipal energy companies, profit sharing (between the company and the customer who benefits from the savings) appears an obvious solution, but is not always supported by regulatory authorities. Here, changes are necessary. In terms of public transport, greater resources need to be provided by central governments; these can, for example, be diverted from road programmes or levied through additional car taxes. Undoubtedly, as the example of the UK has shown, deregulation is not the solution for providing an efficient and low emission public transport system.

One problem is that national governments have paid very little attention to the scope for local action in their climate change strategies. As already mentioned, there is a lack of coordination of activities between national and local levels. Furthermore, there needs be a greater effort to encourage local authorities to formulate climate change strategies, for example by providing grants for the drawing up of emission inventories and strategies. As yet, the local approach to climate protection is too piece-meal, with too few authorities involved, especially in the Southern Member States. Within this context, EU financial support and the activities of ICLEI and the Climate Alliance are important.

CONCLUSIONS

Through the examination of developments in the EU and the case studies of the six Member States, herein we have attempted to highlight the most important political realities that constrain an effective drawing up and implementation of climate change strategies. We believe that recognition of these constraints is an important step in the search for adequate precaution-ary action, but also that this recognition is yet to manifest itself in an effective way at any decision making level in the EU, either within national administrations or on a collective basis.

The study has found that while Germany and the UK are projecting emission reductions, these may be short lived. Elsewhere, emissions are, at best, being stabilized for the year 2000. Although each country has drawn up a strategy and is employing some policy measures, in most cases these are incidental, resulting from decisions taken essentially on economic grounds. While this is in itself a positive development, showing that there are areas of overlap between environmental and economic objectives, these measures are at the same time insufficient to achieve substantial emission reductions. Considering the problems with achieving the targets for the year 2000, efforts are obviously necessary to overcome these constraints if emission reductions post 2000, as would be required if binding protocols were adopted under the FCCC, are to become a reality.

A range of constraints has been identified in this study, but many of these are to some extent part of the same problem, namely the lack of a coherent and comprehensive integration of climate change objectives into economic decision making. Priority continues to be given to traditional objectives such as economic growth and industrial competitiveness, and short term thinking prevails. These objectives are not necessarily incompat-ible with climate change objectives, especially with regard to the possibilities for CO_2 emission reductions related to energy efficiency initia-tives. However, the public, political and especially industrial opposition display a deep reluctance to see any real changes to the status quo across wide sections of society. Also, in some cases economic benefits accrue only in the long term and at the aggregate level, while short term costs occur for certain actors. Other, more specific, constraints identified include various procedural difficulties, current regulatory frameworks for the energy sector, a lack of cooperation between central and local government, low energy prices and industrial lobby influence and the lack of public awareness. This is not to say that such problems are insurmountable. There have been some improvements in understanding and consideration of the climate issue at both a governmental and a societal level, but this progression needs to continue further towards a reconsideration of policy priorities, with a particular emphasis on the participation of all the relevant actors in the policy process. As far as the involvement of the general public is concerned, regional and municipal tiers of government and NGOs can (and generally show a willingness to) play an important role in tackling such issues at a local level. Appropriate dialogue between policy makers and citizens should be pursued more seriously not only as a way of addressing the climate issue but in response to the wider environmental action frameworks, in particular Local Agenda 21. As regards industrial opposition to specific or general measures, continued dialogue is also very important if constructive rather than regressive or confrontational (and therefore unworkable) action is to be taken.

The implications of European integration within the context of the climate change issue, especially with regards to the transport challenge, are likely to endanger any non EU level actions on climate change. The application of the free circulation principle has brought with it a huge increase in freight transport (especially road based), as well as in passenger transport. Differing production costs have meant that imports may be cheaper than home produced goods, primarily because transport costs are low. The apparently myopic pursuit of economic growth through the policy of European integration and cohesion exposes inherent incompatibilities with any aim to reduce greenhouse gas emissions, a situation which, in the opinion of the authors, is an inexcusable evasion of what could be achieved at an appropriately continental scale. Action at the EU level needs to place greater emphasis on ensuring the compatibility of the institution's own policies, especially between those on the economy and environment, in particular climate change objectives. Within this context, the EU should:

- Integrate the IEM proposals with proposals on rational resource planning to remove contradictions and ensure maximum environmental compatibility. This might take the form of setting specific targets in liberalized markets for beneficial options such as CHP and DSM.
- Include climate change emission reductions as a criterion in the allocation of structural funds (ie priority to public transport projects, renewable energy funding). This would be important for supporting regional and local authorities in their activities.
- In the Trans European Network plans, give priority to public transport.

As concerns the crucial aspect of energy efficiency, the most logical area for further EU action is to push for agreement on consumption standards for all major electrical appliances and motor vehicles. Considering the problems encountered with the carbon/energy tax, there are clear feasibility problems with regards to EU level environmental taxes. Taxes do present a useful instrument (although they should not be seen as a panacea) and it is possible that progress could be made with a change of government in the

UK, since the present Conservative administration has been the main objector to the tax.

Overall, therefore, there is a pressing need for an effective EU level climate change strategy, as a number of Member States are unlikely to act unilaterally. Climate change is undoubtedly one of the most complex environmental problems policy makers have to deal with because addressing the issue requires an integrated response in a number of policy areas. In view of the fact that the effects of climate change will almost undoubtedly be on a global scale (even if the exact effect continues to be a matter of dispute) it would seem reasonable to expect that sustained, concerted and coordinated action should be based on as wide a political and geographic range as possible. With most greenhouse gases originating in the industrialized 'Western' capitalist societies, it is appropriate that it is from these regions of the world that the initiatives for change should come. If reasonably stable unions of peoples such as the EU cannot find ways of addressing the climate change issue it is difficult to see how it can be addressed at all.

INDEX

Milestones in Ozone Protection

Elizabeth Cook

On 1st January 1996 the Montreal Protocol banned the production of chlorofluorocarbons (CFCs). The effectiveness of this milestone in global environmental protection offers valuable lessons for government, industry and environmental organizations. This book examines the success story of the CFC phaseout. It shows how the joint efforts of economic incentives, entrepreneurial government activities, corporate leadership and competition, scientific advances and public activism significantly aided the adoption of CFC alternatives.

£14.95 *paperback* ISBN 1 56973 088 1

The Economic Implications of Climate Change in Britain

Edited by Martin Parry and Rachel Duncan

"A useful and reliable book which indicates changes to come in Britain in several vital areas of life" Agroforestry News

Climate change could have a substantial economic impact, especially on coastal states where rising sea levels will be felt most strongly. Within this book leading experts from organizations such as the Inter-Governmental Panel on Climate Change (IPCC) and the UK Climate Change Impacts Review Group have examined alternative scenarios for change and the implications invloved. They look in detail at agriculture and land use, the finance and insurance sector and water supply and management. This is an important issue and this book makes a major contribution to understanding what is at stake.

£14.95 *paperback* ISBN 1 85383 240 5

Valuing Climate Change
The Economics of the Greenhouse Effect

Sam Fankhauser

"Sam Fankhauser has produced the most careful analysis yet of the likely economic damages arising from global warming" from the Foreword by David Pearce

Valuing Climate Change examines the issue of global warming and highlights the urgent need for action to be taken. Fankhauser assesses the costs of a doubling of greenhouse gas (GHG) emissions to be a significant percentage of gross world product; a figure which he then compares to the costs of reducing emissions. In his comparison he examines regional and global estimates of damage, and takes into account the non-climate change benefits of GHG reductions. Fankhauser shows that tougher targets may be needed than those set out in the Framework Convention on Climate Change. He also assesses the optimum policy responses to GHG reduction, the instruments that could be employed for achieving it and the potential for international cooperation in dealing with the problem.

£14.95 *paperback* ISBN 1 85383 237 5

Energy Policy in the Greenhouse

Florentin Krause, Wilfrid Bach and Jon Koomey

"A book [which] *throws into stark relief the mountain still to be climbed before the world community can agree on a credible programme to tackle global warming"* David Thomas, *Financial Times*

Global warming is occurring at an alarming rate. Using computer modelling, *Energy Policy in the Greenhouse* looks at ways of minimizing the risks. Fossil carbon emissions, other trace gases and releases from other sources are taken into account. The authors demonstrate the need to return to a rate of forest carbon storage equal to that of the mid-1980s and they demonstrate the need for a global budget for cumulative releases up to the year 2100. These budgets look at issues of international equity and the ways of moving towards a binding agreement. The price of failure to control GHG emissions may be uncertain but it will be more than anyone can afford. This book provides an agenda for advance.

£17.50 *paperback* ISBN 1 85383 081 X
£29.95 *hardback* ISBN 1 85383 080 1

Earthscan Publications Ltd
http://www.earthscan.co.uk